THE AMERICAN MARITIME LIBRARY

VOLUME X

EDWARD DAVIS THAYER
COLLECTION

STARBUCK'S NECK, EDGARTOWN

Down on T Wharf

The Boston Fisheries as Seen Through the Photographs of Henry D. Fisher

By ANDREW W. GERMAN

MYSTIC SEAPORT MUSEUM, INC.

Mystic, Connecticut

FRONTISPIECE:
T Wharf, south basin, 1911

FIRST PRINTING, 1982
Copyright © 1982 by Mystic Seaport Museum, Incorporated

Cataloging in Publication Data

German, Andrew W. 1950–
    Down on T Wharf: the Boston fisheries as seen through the
photographs of Henry D. Fisher, by Andrew W. German.
Mystic, Mystic Seaport Museum, Inc., 1982.
    xvi, 160 p., illus., 28 cm. (American Maritime Library, no. 10)

    1. Fisheries—Boston—Pictorial works.
2. Shipbuilding—Essex, Mass.—Pictorial works.
3. Fishing Boats—Pictorial works.
I. Fisher, Henry D. 1882–1961. II. Title.
SH 221.M3G4
ISBN: 0-913372-26-9

Designed by Klaus Gemming, New Haven, Connecticut
Composed by Finn Typographic Service, Inc., Stamford, Connecticut
Printed by Eastern Press, Inc., New Haven, Connecticut
Bound by Robert Burlen & Son, Inc., Hingham, Massachusetts

Printed in the United States of America

CONTENTS

FOREWORD

THERE WAS A TIME, not so long ago, that a reader looking for good books in maritime history had little more to choose from than hefty academic tomes designed for graduate students, flossy collections of illustrations suitable for coffee tables, or salty romances (what Robert G. Albion used to call "bust-and-frigate" books) seemingly written as grade B movie scripts. Then in the fall of 1967, after a series of conferences with Willard A. Lockwood, Director of Wesleyan University Press, Waldo C. M. Johnston, Director of Mystic Seaport, proposed that the Museum and Wesleyan join to create the American Maritime Library, an on-going series of books on all aspects of American maritime history except strictly naval subjects. An editorial advisory committee was appointed to the task of selecting manuscripts in a wide range of categories—historical narratives, biographies, journals, and monographs—whatever met the committee's standards of significance, accuracy, and readability. The series was designed primarily to meet the interests of the Museum's 15,000 members, who, because of their knowledge of and commitment to America's maritime heritage, were not easily satisfied by the usual potboilers. If the series reached others with similar taste, so much the better.

With William A. Baker, Marion Brewington, and Willard Wallace on the editorial committee, meetings to discuss manuscript submissions were never dull. It is a wonder they agreed to publish anything! Whatever might fulfill Marion's keen sense of what was an important contribution to maritime history usually came acropper under Bill's sharp eye for nautical detail. It was then a virtual certainty that the few survivors would fail to meet Will's high standards of good writing.

Nevertheless, after a year of screening, the committee recommended for the Library's first volume Michael J. Mjelde's study of Donald McKay's last ship, *Glory of the Seas.* Meanwhile, subsequent volumes were in the pipeline. John F. Leavitt shared his experiences in coastal sail through his magnificent album-sized work, *Wake of the Coasters,* and Curtis Dahl edited the whaling journal of his great-grandfather, Ben-Ezra Stiles Ely, under the title *There She Blows.* In 1972 the series produced two more works. The first was a remarkable study of three generations of maritime entrepreneurship, *The Mallorys of Mystic* by James P. Baughman. Three members of the editorial board, Bill Baker, Marion Brewington, and Ben Labaree, joined with Robert G. Albion to write the history of *New England and the Sea.* In the following year, Horace Beck's collection of maritime beliefs and stories was published under the title *Folklore and the Sea,* and in 1974 appeared the fascinating study of *Oystering from New York to Boston,* by John M. Kochiss.

Like so many other publishing programs, the American Maritime Library encountered hard times during the latter half of the 1970s. The venture was confronted by the catch-22 of modern publishing; no new money, no new titles; no new titles, no new money. In 1977, however, the sponsors were able to issue James Garitee's authoritative study of privateering in Baltimore entitled *The Republic's Private Navy.* But soon afterwards, Wesleyan University Press underwent a major reorganization, merging its operations with Columbia University Press. Joint sponsorship of the American Maritime Library ended, and Mystic Seaport Museum was on its own.

Under the able direction of Jerry Morris the

Museum had been expanding its own list of publications for several years. He suggested, and Director Revell Carr and the trustees agreed, that the Museum should continue the American Maritime Library series under its own imprint alone. The editorial board was reorganized and given wider responsibilities to include oversight of all historical works published by the Museum, and the first book under the new arrangement appeared in 1980. Briton Cooper Busch's edition of Captain Joseph J. Fuller's journal, entitled *Master of Desolation,* told the exciting story of the little-known sea elephant trade off Kerguelen Island in the southern Indian Ocean. Publication of that volume established what the editors anticipate will be a regular schedule of an American Maritime Library book every other year, not so prolific a pace as in the early 1970s but perhaps more realistic in terms of present conditions in publishing.

Given that the Museum's primary purpose is to preserve America's maritime past, it would be difficult to imagine a book that better fulfills the goals of the institution than Andrew German's *Down on T Wharf.* When the editorial committee viewed the photographs and read the accompanying text they agreed with rare unanimity that this was just the kind of project for which the series was initiated.

Down on T Wharf is much more than an assemblage of old photographs, and Mr. German has not been content merely to select pictures and tack on captions in the mistaken impression that whatever a photograph has to say it can say for itself.

The author's textual material is of two sorts—short sections introducing each of his several chapters, and generous commentaries (I refuse to call them captions!) accompanying each illustration. Taken together the introductory paragraphs offer a capsule history of the Boston fishing industry at the beginning of the twentieth century. Disclaimers in his pref-

ace notwithstanding, Mr. German has managed in these eight introductions to provide the essential background so that even the relatively uninitiated reader can greatly profit from the text itself.

Generous footnotes and the author's annotated bibliography add immeasurably to the value of this work. None of the more recent books on New England fishing has had adequate references, and the old standbys like Innis and McFarland are quite out of date.

One more comment should be made about *Down on T Wharf:* the fact that it was written by a young member of the Museum's staff. Such an occurrence at one of Europe's historical museums would not be at all remarkable; but in America scholarship has played second fiddle to visitor services in budgetary considerations at most historical museums. There have always been special exceptions of course, but usually these opportunities fall only to the older, established members of the profession who have "earned" such privileges.

Perhaps the publication of Mr. German's book will stimulate administrators and trustees to consider more seriously the role scholarship plays in the life of the museum. Research and writing together comprise one of the most rewarding of all museum activities. They nurture the minds of those who undertake them and add to our knowledge and understanding of the collections under our care. And finally, publication of the results of research represents a double outreach for the museum, beyond the geographical limits imposed by its location and beyond the intellectual limits imposed by its necessary commitment to the general public. In expanding its publications, a museum has the opportunity to expand itself as an institution.

Benjamin W. Labaree, *Chairman*
Mystic Seaport Museum
Editorial Advisory Committee

viii

PREFACE

THIS BOOK has been a labor of love for me since the day I came across a large cache of negatives in my grandfather's house in New Haven. Stored in brittle, crumbling envelopes, they had lain largely undisturbed since the 1920s. Unknown since the death of my grandfather, Henry Donald Fisher, in 1961, they came to light twelve years later when, upon the death of his wife, we began sorting through the contents of their home. Opening a drawer full of negatives, I pulled out a view of a New Bedford whaler, and realized at once that I had uncovered a remarkable treasure.

The collection contained several thousand negatives, covering the years from 1905 into the 1930s. They formed quite a complete record of Henry Fisher's family and friends, his varied interests, and his travels. But it was the nearly 800 maritime views that claimed my attention. They offered a mixture of yachts, commercial vessels, and naval vessels, sail and steam, wharf scenes and views taken on the water. Most were taken around Boston and Massachusetts Bay, or near Philadelphia.

My grandfather's apparent interest in fishermen and fishing vessels curiously paralleled my own. While studying his nearly 150 fishing views, I realized that they were quite unusual in that they presented one man's private, amateur view of the fishing fleet in Boston Harbor. A fragile view it is too, as the images are fixed on perilously unstable nitrate-based negatives.

The views are perhaps most significant in providing a detailed, though random, record of the shoreside components of a wide range of fishery activities during an important transitional period. As the United States emerged from the nineteenth century, the technology of the fisheries began to change rapidly.

These photographs record changes in fishing schooner design, the proliferation of internal combustion engines in the fishing fleet, the appearance of the otter trawl method of fishing (which has since superseded the other methods pictured here), and the modernization of the shoreside aspects of the fisheries. All of these changes became well established in the first twenty years of the century, the period that embraces these views.

The views also show the fishermen quite clearly. The schooner fishermen, who would soon have to choose between new techniques and their accustomed way of life, the steam trawlermen who had already accepted the new way, and the newly-established Italian immigrant boat fishermen all appear candidly in their element.

There were several schemes of organization that might have been applied to the 117 views selected for publication. As I studied them, I found that they fell most logically and clearly into an arrangement depicting the sequence of events on T Wharf, and the aspects of change then developing. After composing a photographic tour of T Wharf and its successor, the South Boston fish pier, I have arranged the views in chapters on the schooner fleet showing construction, outfitting, departures and arrivals, and the discharge of fish. The changes in the fleet are acknowledged in chapters on the passing of the schooners, the proliferation of the steam trawlers, and the Italian boat fishermen of T Wharf.

To elaborate on the photographs, I have organized long descriptive legends to verify visible details and offer some background information on the vessels and activities depicted. Drawing largely from government reports, periodical articles, and newspaper items, a great effort has been made to provide caption material contemporary with the photographs, thus enhancing their validity and immediacy.

It should be remembered that the legends are designed to serve the photographs and do not comprise a full history of the fisheries. Much background information is assumed on the reader's part. Some topics, notably the long and involved diplomatic history of international fishing rights, are barely mentioned. A glossary and a list of sources have been included to offer the reader some background information and directions for further reading.

I have tried not to overemphasize the hazards of the fisheries, which are probably exceeded only by those of the mining industry. Between 1908 and 1913 the mortality rate in the fleet varied from roughly one in 85 to one in 230. It was a fact the fishermen merely had to accept, perhaps with religion, perhaps with a fatalistic viewpoint, and often with a devil-may-care humor.

I have also attempted to date the photographs as exactly as possible. The earliest view included was taken in 1907, the majority date between 1908 and 1915, and the last was taken about 1922. Through 1909, Henry Fisher dated each negative, noting the time, light conditions, and exposure data. Those taken thereafter are simply identified with film roll and exposure numbers. I have worked out at least approximate dates for these, using negative numbers and seasonal clues, and, most importantly, matching identifiable vessels with newspaper listings of vessel arrivals. With patience, it was possible to date many views using the *Boston Post* and *Gloucester Daily Times*, especially when more than one vessel could be identified in a view, or in a series of views taken on the same day. Sometimes a vessel without a visible name could be identified by determining a date and comparing the vessel's characteristics with those of the vessels in port that day. Some identifications are

Henry D. Fisher, 1908

listed as probable because they could not be absolutely verified. Gordon W. Thomas was extremely helpful in corroborating identifications.

Finally, it should be noted that the photographs have been cropped somewhat. Usually, this has meant excluding empty water or incomplete detail on the margins of a view, but in one case a long, thin panorama negative has been cropped extensively. Fisher's own photographs reveal that he cropped when printing. Without the variety of close-up and telephoto lenses available today, photographers often had to include unwanted detail in their exposures and crop it out during printing.

Born in August 1882, Henry Fisher was the son of Andrew O. Fisher, an insurance agent, and Elmira Bevan, daughter of an old-line farming family in the Ardmore, Pennsylvania, area. He grew up with the spread of urban life and the diffusion of amateur photography, which blossomed when George Eastman marketed the first Kodak camera in 1888. There is no evidence, however, that Fisher picked up a camera during his childhood in Philadelphia and Haverford, or during his engineering studies at the University of Pennsylvania. Graduating in 1904, with a degree in mechanical engineering, he took an apprenticeship at an engineering firm and then worked as a draftsman at a boiler plant in Philadelphia. In 1905 he started to use a simple Kodak Bullseye camera, capturing images of his family, the woodlands of Haverford, scenes around Philadelphia, and summer vacations at the Jersey shore. His first maritime views were made on the Delaware River and at Egg Harbor, New Jersey.

In 1908, he moved to Boston to work for the Arthur D. Little Company as a field engineer. He spent a pleasant year, boarding on Rutland Street, sailing Boston Harbor in a sloop he shared with friends at the Quincy Yacht Club, and becoming acquainted with Boston. Now with a more sophisticated camera, he began to roam the city.

He also began to visit T Wharf occasionally, to photograph the fishing fleet. Like the American cowboy, the fisherman had acquired a romantic mantle in an age of more prosaic pursuits. Likened to a "modern Norseman," the brave, independent fisherman was considered by some a "modern industrial hero." James B. Connolly spread the fisherman's renown during the decade after 1900 through his nationally-known short stories. By 1909, Henry Fisher owned four volumes of Connolly's stories, and was likely influenced by them in his choice of photographs.

In March 1909, he moved to Pittsburgh to work for the U.S. Glass Company. There he used his camera in the city and as he travelled the Midwest. He also found boating opportunities in a canoe club on the Allegheny River and during summer vacations on Lake Keuka, New York.

After a year in Pittsburgh, he returned to Boston to work as a field engineer for the Fuel Testing Company. As he had in the Midwest, he carried his camera during his travels about the region, inspecting coal-fired boilers from Rhode Island to Maine. On lunch hours, he often walked around Boston, and on weekends took a trolley to the suburbs, or a steamer to Gloucester or Provincetown, photographing street scenes, haying on the Lynn marshes, and woodland and shore scenes.

Married to Edna Pursell of New Jersey in 1911, he settled in Newton Highlands. Within three years he had two children, and in 1915 he moved to the Bailey Meter Company. Gradually, his photography, particularly of maritime subjects, dropped off. With a family and job-related travel and responsibilities, Fisher, now in his mid-thirties, had less time for the solitary pursuit of photography, except of his children and during his travels.

Then, in 1920 he took his family to New Haven, joining the New Haven Pulp and Board Company. Except for a year at a coal mine in Pennsylvania, 1922–23, he spent the rest of his career rising into the management of the New Haven Pulp and Board Company. He also wrote many technical articles on coal combustion and developed an interest in genealogy, and in the work of Mystic Seaport.

Until his death in August 1961, Fisher maintained his engineer's precise interest in photography, keeping up with improvements in equipment and technology. In his approach he was what George Eastman had called a "true amateur," with an eye for composition, an interest in the interplay of light and shadow (represented by numerous views of breaking waves and snowy woods), a concern for the mundane activities of the day, and a fascination with the process of photography. And though his photographs were mainly intended for his own amusement, he did exhibit the view entitled "Up Jibs" (page 67) at a photography show sponsored by the New York branch of John Wanamaker's Department Store in 1913.

Until about 1915, he did his own developing and printing, mixing his own chemicals. Some negatives are flawed with fixer stains or other problems from his primitive bathroom darkroom. Other negatives are flawed because the nitrate base began to deteriorate in storage. Others are imperfect because he attempted exposures in imperfect light or inclement weather. In all cases, he would have considered the views unacceptable, but after seventy years the historical value of the views outweighs the flaws.

Were it not for Eastman's development of flexible gelatine sheet and roll film in the 1880s, these spontaneous, "instantaneous" views could not have been taken. Fisher's Kodak No. 2 Bulls-eye, Kodak Snapshot, Panoram Kodak, and Kodak folding 3A cameras were easy to operate and often produced remarkably clear images, up to $3\frac{1}{4}$ x $5\frac{5}{16}$ inches. But many of the most immediate views were made with a diminutive "Ernemann Detectiv" camera, made in Dresden. In 1912 he apparently bought this 6 x 3 x 1½-inch folding camera with an 80mm lens, and thereafter felt more free about snapping individuals and wharf activity. This precursor of the 35mm camera produced $1\frac{9}{16}$ x $2\frac{7}{16}$-inch negatives. Their only drawback is coarse grain and a lack of contrast.

Henry Fisher's vision will live on in these photographs, and to make them available they have been deposited at Mystic Seaport Museum. Over 800 negatives have been copied and are available for research.

I could not have completed this work without the consultation, criticism, assistance, and encouragement of numerous individuals. My mother, Dorothea Fisher German, and her brother, Henry Donald Fisher, Jr., were both pleased to have their father's work attain historical significance. Gordon W. Thomas, son of a noted fishing captain of Gloucester and possibly the living expert on Gloucester's schooner fleet, gave me the benefit of his immense knowledge and encouraged this view of Boston's fisheries from the beginning. Dana Story, whose father built many of the schooners pictured here, carries on the tradition, and offered valuable assistance. Both Charlie Sayle of Nantucket and Francis "Biff" Bowker of Mystic cast their practiced sailors' eyes over the work and made suggestions, as did Professor Benjamin W. Labaree, Seaport Curator Benjamin A. G. Fuller, and the late William A. Baker. The gang from the Gloucester Fishermen's Institute imparted a sense of the dory fisherman's life afloat during their visits to Mystic Seaport.

As sources on the fisheries are generally obscure, I became dependent on several excellent libraries, spending many hours at the G. W. Blunt White Library at Mystic Seaport Museum and the Boston Public Library, as well as visiting the Sawyer Free Public Library at Gloucester, the James Duncan Phillips Library of the Essex Institute in Salem, and the libraries at the Cape Ann Historical Society in Gloucester and the Bostonian Society in Boston. The Baker Library of the Harvard University Business School and the Massachusetts State Archives were also generous in providing material.

At Mystic Seaport Museum I am grateful for the help of Registrar Philip L. Budlong, Rodi Hamilton, Richard Malley, and William N. Peterson of the Curatorial Department; Virginia Allen, Lydia Frank, Anne Goodrich, Lisa Halttunen, and Douglas Stein of the G. W. Blunt White Library; Kenneth Mahler, Mary Anne Stets, and Claire White-Peterson of the Photo Lab; and Glenn Gordinier. Bill Gill drew the maps and Paul Gaj drew the fish for the appendix. Finally, I must thank my wife Amy, who has supported me through this project for so long.

AN ODOR OF FISH rushes upon you as
you saunter along what the fishermen call
"the Avenue," and when you are a little
past State Street, on your way to the
South Ferry, the odor strikes you with
a good, smart whack. A maze of little
masts; a fleet of tiny vessels; men,
some of them swarthy as Moslems; men
hauling on ropes or drawing carts laden
with the food which the sea gives forth—
this is T Wharf.

Charles A. Campbell,
"The *Mermaid's* Nursling,"
New England Magazine 41 (1909).

1 T Wharf

THE STORY of T Wharf here is but a part of the larger story of the fisheries in Boston. It is a story which can be traced from 1835, when the first wholesale fresh fish dealer established himself on the waterfront, and continues to the present day.

The beginnings of T Wharf actually extend back to the 1720s when it was known as Minot's T. An adjunct of Long Wharf, it was attached to that wharf by remains of the 1673 defensive barricado and formed a "T" through this connection. Its greatest physical growth occurred in the late 1820s, as all of Boston's wharves underwent development, reaching out for the port's growing maritime commerce. Through the mid-nineteenth century T Wharf remained a commercial spur of Long Wharf, serving coastal packet traffic, especially that to Maine and the Maritime Provinces. The Atlantic Avenue project to improve waterfront access, which required much waterfront filling, was conceived in 1868 and begun in 1869. Partially following the line of the ancient barricado, Atlantic Avenue effectively reduced T Wharf's length but improved its access. The wharf had experienced a century and a half of commercial booms and busts, and was in a graceful decline when the fishing industry began to eye it in 1882.[1]

Boston's first wholesale fresh fish dealership, Holbrook, Smith & Co., opened a store on Long Wharf in 1835. Up until that time, individual fishermen had been able to supply the city's demand for fresh fish. As markets expanded within the city and beyond, the number of dealers grew, particularly after 1860. Seeking a convenient location on the waterfront, dealers settled on Commercial Wharf, first in wooden sheds and then, by 1870, in the granite block of stores built originally for merchants dealing with the East Indies and other lucrative markets.[2]

By 1880, the fish dealers were looking for a new, consolidated location, and nearby T Wharf appeared suitable. Some dealers were already located there, but, except for those in the brick building at the inner end of the wharf, they were accommodated in rudimentary wooden sheds. In 1882–83, a long row of wooden stores was built the length of the wharf. The thirty fish dealers desiring to move to T Wharf bid on the stores in the new building and moved to their new quarters in 1884. The T Wharf Fish Market Association, formed by the dealers, paid yearly rent and taxes, plus 10 percent of the cost of the building.[3]

One inconvenience of T Wharf was the limited water rights included in the lease taken by dealers on the wharf. Only a section of the north basin thirty feet wide was included with the wharf, so the Fish Market Association was forced to pay $5,000 a year to the owners of Commercial Wharf to allow fishing vessels to lie abreast of each other on the north side of T Wharf. Otherwise, the stores were modern, land access was good, and if the wharf did get congested, it at least had 1,200 feet of frontage, not including the outer end where towboats often lay.[4]

T Wharf was occupied on the basis of a

ten-year lease, and this lease was renewed in 1894. However, by 1904 dealers had spilled out onto Atlantic Avenue, and T Wharf was obviously overcrowded, congested, and gaining the reputation of being unsanitary. Unable to find an alternative location, the T Wharf Fish Market Association renewed their lease once more. Four years later they formed the New England Fish Exchange to regulate the fresh fish business and, in the spirit of cooperation, formed the Boston Fish Market Corporation in 1910. This corporation, working with the Commonwealth of Massachusetts, developed a 1,200-foot pier on the flats of South Boston. Massachusetts constructed the pier, 1,200 feet long and 300 feet wide, of fill faced with granite. The pier was begun in the fall of 1910, and completed in the spring of 1913, at a cost of about one million dollars. The Boston Fish Market Corporation then spent another million constructing two blocks of fireproof brick stores 50 feet wide and 750 feet long, an administration building at the end of the pier, and other ancillary buildings. The Commonwealth Ice & Cold Storage Company put up the world's largest freezer and ice plant at the inner end of the pier.[5]

In a ceremony and procession that combined nostalgia for the traditions and successes represented by T Wharf with ambitions for the claim to greatest fish mart in the world represented by the new pier, dealers moved their businesses from T Wharf to South Boston on 28 March 1914. The pier opened for trading on Monday, 30 March 1914, and business continues there today. The dealers achieved their aim of becoming the world's largest fish marketing location, in fact holding that claim during the First World War. Through the 1920s and '30s they had the largest market in the Western Hemisphere, if not the world.

But it took great effort to attain that position. Appendix 2 presents the firms that moved from T Wharf to South Boston. Within these brief biographies can be glimpsed the backgrounds of some of the aggressive and opportunistic merchants who built up a large and diverse industry in a brief time.

When the industry relocated to T Wharf in 1884, Boston had 51 schooners (13 percent of Gloucester's 388) and 876 fishermen (15 percent of Gloucester's 5,778). Landings of fish at Boston in 1887 totaled 18,514,086 pounds worth $383,973, only 18 percent of the tonnage and 13 percent of the value of Gloucester's 101,511,113 pounds worth $2,795,229.[6]

Over the next ten years, the market for fresh fish developed by Boston dealers brought fish in from around the country, and attracted an increasing proportion of the produce of Gloucester's vessel fisheries. Gloucester's wholesalers were largely committed to the salt cod fishery, the halibut fishery, and the salt and fresh mackerel fishery. The much broader demand in Boston brought Gloucester vessels into Boston until, by 1897, Boston received 62,903,558 pounds worth $1,230,044. That was 93 percent of the tonnage and 72 percent of the value of Gloucester's landings, and a more than 300 percent increase in ten years.[7]

Gloucester's dealers and vessel owners eyed the situation with concern, thinking of revenues lost to Boston fish dealers and outfitters. In 1897, they created the Gloucester Fresh Fish Company to attract Gloucester vessels back home. The strategy worked and Boston's landings dropped by 8 million pounds (12 percent) while Gloucester's rose by 21 million pounds (31 percent).[8]

Boston had never developed a sizable fleet, having 60 schooners in 1898 (an increase of 9 in thirteen years), as compared to about 350 vessels out of Gloucester. Because Boston had concentrated on its marketing network and depended on Gloucester's fleet, it took five years for landings in Boston to climb close again to those in Gloucester. In 1903 Boston received just over 80 million pounds of fish and Gloucester just under that figure.[9]

Another five years of rough parity ensued until, in 1908, Boston landed 95,659,680 pounds worth $2,565,010, compared to Gloucester's 85,805,567 worth $2,064,415. Gloucester vessels were still bringing their fares into Boston, but Boston had built up its fleet to 105 fishing vessels (including 9 power vessels) while Gloucester's fleet had decreased slightly, to about 275 (including 62 power vessels).[10]

During that time, a strong effort was made to consolidate the Boston fish dealers into a combination that would control both fresh and salt fish, set the price of fish according to supply and demand rather than bid, and finance a revolution in the fisheries by sponsoring large-scale development of otter trawling. The National Fisheries Company, as it was called, was capitalized for $5,000,000 by the association of fourteen wholesale firms, which turned their assets into stock. Along with the dealers themselves, the New England Fish Company, which operated fishing steamers on the West Coast, and the Bay State Fishing Company, which had inaugurated New England otter trawling in 1905, joined in the combination. In an era of trusts, this potential fishing trust had its supporters when it was founded in 1906, but in the end opponents among the wholesalers and advocates of the line fishermen prevented its establishment.[11]

In 1908, a successful year despite the national recession, Boston dealers attempted a cooperative venture in fisheries management called the Boston Fish Exchange. The Exchange instituted a regulated auction form of sale for fish as they arrived, and held both fishermen and dealers to their word on fish quality and price. These revised methods, developed and supported by both wholesalers and fishing masters, were seen as the greatest occurrence since the opening of T Wharf as a fish market.[12]

At the same time, dealers had to contend with sanitary conditions on the wharf that the health commissioner described as the unhealthiest in Boston, due to decaying fish refuse on the wharf and in the basins on either side. An outlay of $80,000 by dealers, for a sewer on the wharf, helped alleviate the problem, but all agreed that a new location was necessary.[13]

The Boston Fish Market Corporation of 1910 continued the progressive movement of the Boston fisheries. The modern, efficient facility that it sponsored in South Boston was well calculated to assure Boston's position as America's leading fish market. In 1914, the year that the new fish pier opened, 92,344,192 pounds of fish, worth $2,613,987 were discharged at Boston, while Gloucester received 70,245,028 pounds, worth $1,781,043. In that year it was thought that Boston was the world's second largest fishing port, ranking slightly behind Grimsby, England, in tonnage landed, but slightly ahead in value of fishery products.[14]

Boston's position as a transportation center contributed much to the success of its fish dealers. Regular steamboat routes eastward to Maine and the Maritimes, through which came fish supplementary to those landed by fishing vessels, and along which were sent retail fish products, formed one part of the network. In the other direction, water routes to New York and southward were used, but the railroad was the primary transport artery for the fish business. On it arrived fish from the West Coast and Gulf fisheries, consigned to Boston wholesalers, and on it departed half the fish that passed across T Wharf, bound for retailers and intermediaries throughout the eastern half of the nation. Improved communication by telephone, improved transportation, improved refrigeration methods, and fish prices competitive with beef prices helped spread the demand for fish into the interior.

So did the great influx of immigrants who rushed to America. In 1914, the *Fishing Gazette* optimistically reported that 1,300,000 immigrants had entered the United States in 1913, and most of them were fish eaters. Fish was cheap, fish was customary food for many Europeans, and fish was required food for the many Catholic immigrants, who made Lent a fruitful time for the fisheries. Perhaps it was the large percentage of immigrants in the textile industry of the Merrimac River valley that influenced the investment of textile money to establish the Atlantic Maritime Company. Atlantic Maritime owned thirteen vessels by 1906, and landed its fish at T Wharf in winter and Gloucester in summer.[15]

The widespread demand for fish returned a bounty to dealers and fishermen of T Wharf. As operations ceased there and dealers packed to move to their new, modern quarters in South Boston, it was revealed that in the preceding five years, $46,000,000 had been sent to the wharf from the bank to pay for cargos landed, carried inconspicuously in paper bags. The pattern of increased demand, increased technological improvement, and increased profit continued with the opening of the new fish pier. Landings of fish in Boston rose from 95 million pounds in 1908 to 124 million pounds in 1923, to 285 million pounds in 1930. Also, between 1914 and 1930 the number of fishery-related jobs in Boston doubled and wages for these jobs tripled.[16]

Even when the fish dealers abandoned T Wharf for South Boston, the Italian boat fishermen continued to use the old facility. Part of the Portuguese fleet remained at T Wharf also, and most schooner fishermen preferred laying up between voyages in the shelter of T Wharf rather than on the exposed flats of South Boston. Ties to the fisheries remained when, in 1917, the Quincy Market Cold Storage Company built a large freezer, partially on T Wharf and partially on piles in the south basin between T and Long wharves.[17]

Like neighboring wharves, T Wharf lost most of its maritime connections as Boston turned from the sea; and the passenger and bulk cargo marine commerce remaining in Boston proved unsuited to the crowded nine-

teenth-century wharves. But the fishy atmosphere of T Wharf only added to its attractiveness in the eyes of a contingent of Boston artists and writers. Through the 1920s and '30s a colorful artistic presence monopolized T Wharf, commemorated by Z. William Hauk in *T Wharf Notes and Sketches*.

However, without maintenance, the rambling wooden block on the wharf began to sag and the wharf to crumble. On the then-blighted waterfront it was merely routine to condemn the useless structure. T Wharf was demolished in the early 1960s, eighty years after entering its heyday as the focus of the fishing industry, and fifty years after being abandoned by that industry.

Looking Out T Wharf on a Typically Busy Day, 1911

ON THE RIGHT, two fishermen in sea boots, one wearing a natural-colored oilskin suit, head uptown. The young man in overalls, tie, and cap probably works on the wharf. The stocky man in a jersey, carrying a basket, is one of the Italian boat fishermen.

On T Wharf, schooner fishermen, steam trawlermen, boat fishermen, draymen, icemen, handcart men, fish merchants and their employees, and spectators mingled in a hurried tangle amid the odors of fish offal, horse manure, tar, and the salt breeze. Shortly before businesses relocated to the new fish pier in 1914 the *Fishing Gazette* reported that, while the new facility would have a haddock weathervane, T Wharf had never needed a weathervane since the odor of the wind accurately indicated its direction.[18]

T Wharf was also a noisy place, with the yelling and joking of the crowds, the rumble and clatter of horses' hooves and iron wagon tires on the wooden and cobblestone wharf, and the calling of steam tug whistles.

On the right is the establishment of C. J. Whitman & Co., dealers in fresh fish, founded in 1880. C. J. Whitman went into business at age nineteen.

Looking out T Wharf

By the time of his death at forty-nine in 1910, he was vice-president of the Boston Fish Bureau and treasurer of the New England Fish Exchange.[19]

T Wharf from the Eastern Packet Pier, Summer 1911

THIS VIEW from the elevated railroad along Atlantic Avenue looks across the north basin toward the harbor islands in the far distance. Two Italian boat fishermen motor into the north basin past an assortment of schooners. The aftermost schooner, with a man aloft, appears to be getting under way, assisted by a Ross Company towboat. Up forward, one schooner has a dory sail set.

Through the mid-nineteenth century, the Eastern Packet Pier extended from Commercial Street out to the irregular roof line just left of the chimney in this view. Filling associated with the Atlantic Avenue project left only the tip of the Eastern Packet Pier in the north dock.

In the immediate foreground is the establishment of the Atlantic & Pacific Fish Co., organized in 1907 by Francis J. O'Hara, Sr. and Jr. Unable to locate on crowded T Wharf, the company settled on Atlantic Avenue, with water access through a shed on the Eastern Packet Pier.

The water rights to all of the north basin, except thirty feet along T Wharf, were controlled by the managers of Commercial Wharf, to the left. The outboard row of schooners along T Wharf could lie there because the T Wharf Fish Market Association paid $5,000 a year to Commercial Wharf.[20]

Two Views of Commercial Wharf and the Eastern Packet Pier from T Wharf, Winter 1910–11

THE FIRST VIEW shows the Commercial Wharf block and the Plant Line (Canada, Atlantic & Plant Steamship Co.) sheds on the end of the wharf. The schooner *Benj. F. Phillips* and a lighter lie at Commercial Wharf. The small single dory vessel on the left is at the Eastern Packet Pier at the head of the north basin. Judging from her white gaffs, booms, and mastheads, she is a small Portuguese schooner. On the right, the little towboat *Valora,* built at Richmond, Maine, in 1878, lies alongside another small schooner at T Wharf.

The second view was taken farther out on T Wharf. Looking back, the *Phillips* is on the right and a Gloucesterman lies at T Wharf on the left. At the head of the basin, where the Eastern Packet Pier juts out, are the small schooner and a schooner-boat, with their heads to the angular pier and a waterboat or gasolineboat alongside.

Above the row of fish dealers' stores at the head of the basin, to the right of the sign announcing O'Brien & Co., Fresh Fish, is the Mercantile Block along Commercial Street. Before the Atlantic Ave-

T Wharf from the Eastern Packet Pier

Commercial Wharf

nue project filled in that area of the waterfront, the Mercantile Block was on Mercantile Wharf. Many chandlers and sailmakers who served the vessels on the waterfront were located here, even after the waterfront moved so far from their establishments.

At the end of the block is the big Quincy Market Cold Storage and Warehouse Company freezer. Before the end of the nineteenth century the technology of mechanical cooling was well developed and freezers like this one began to appear on New England waterfronts. These freezers allowed fish dealers to store fresh fish to meet market demands, and for the first time allowed bait dealers to stockpile precious bait to get through periodic shortages.

Commercial Wharf, dating from the 1830s, was once the province of merchants in the East Indian, South American, and European trades. After the mid-century decline of Boston's overseas trade, the fishing industry gravitated to this imposing wharf, only to move to T Wharf in 1884. Obviously, by 1910 maritime entrepreneurs on Commercial Wharf had been replaced by packers and grease and oil dealers, though the signs may advertise the establishment of Thomas A. Cromwell, a schooner owner and outfitter, who operated a chandlery and grease and oil dealership at 39 and 40 Commercial Wharf. The Plant Line to Halifax located in the sheds on the wharf after the line was established in 1888.[21]

The *Benj. F. Phillips* was an Arthur Binney design, built by A. D. Story in 1901 for Benjamin F. Phillips & Co. of Boston. She was reported to be a fast, easily-handled vessel. The *Phillips* was sold to owners in Fortune, Newfoundland, in June 1911 for about $7,500.[22]

Benjamin F. Phillips was one of two brothers who began as fishermen in Swampscott. Deciding that selling fish wholesale in Boston was more profitable than supplying wholesalers, they established themselves on Commercial Street, moving to T Wharf in 1884. When his brother died in 1892, Benjamin gave his name to the firm, and it continued under his name after his death in 1896. The company was well known in New York City for the quality of its fresh cod.[23]

T Wharf South Basin, Sunday, 23 July 1911

ALL IS QUIET and the fish carts are lined up along T Wharf as market schooners begin arriving to auction their fish on Monday morning. The bell in the cupola will announce the beginning of the fish auction at 7:00 in the morning. In the foreground, a few Italian boat fishermen work aboard their colorful craft.

At Long Wharf, the Dominion Atlantic Railroad steamship *Prince George*, an 1898 product of Hull, England, has steam up and the "blue peter" flying at her fore gaff. She will soon depart for Yarmouth, Nova Scotia. The Boston to Yarmouth steamer, the 'Bear boat' as it was known, delivered many a young Nova Scotia fisherman on the Boston waterfront to find his first site on an American schooner; and many a successful fisherman took the Bear boat home to visit with money in his pocket. Steamers on this run also delivered provincial lobsters, salmon, trout, and processed fish to T Wharf fish dealers.

Along T Wharf lie the knockabout *W. M. Goodspeed* and the plumb-stem schooner *George Parker*. The *Goodspeed* has arrived this morning with 50,000 pounds of haddock, 10,000 of cod, and 4,000 of hake.[24]

The plumb-stem design, borrowed from pilot schooners, was popular for a time in the 1880s. The *George Parker* and *Annie M. Parker* were the last of the type built for the Massachusetts fisheries. Both were constructed in 1901 by A. D. Story, on the model of Denison J. Lawlor's *Susan R. Stone* of 1888. The *George Parker* went mackerel seining and dory trawling out of Boston until 1904, and then fished out of Gloucester. In March 1912 she was purchased by Newfoundland owners and foundered three months later.[25]

The *W. M. Goodspeed* was designed by Thomas McManus and launched by John Bishop at Gloucester on 1 June 1908. She was one of nine knockabout schooners built for the Boston fleet that year, and served as the model for the *Victor & Ethan, Alice,* and *Georgia*.[26]

The *Goodspeed* pleased both her owner, William M. Goodspeed, and her captain, George Perry, with her handling ability. Admittedly, she did take

South basin

South side of T Wharf

a little getting used to. Making a trip a week during the summer of 1908, she twice snapped a topmast while driving in from the fishing grounds, and once misstayed and ran aground while coming up the harbor. But, under Captain Perry she was a consistent producer. Her $29,000 stock in 1908 was fifth best among the market schooners, and her $28,000 stock in 1909 ranked ninth, as did her $33,332 stock in 1910. To keep pace with the others in the fleet, the *Goodspeed* received a seventy-five-horsepower gasoline engine late in 1913.[27]

The *Goodspeed*'s fishermen faced the common hazards of their calling, which could arise, for example, in the unexpected form of a kerosene torch bursting to burn a fisherman's face as he baited his trawls at night. And, like most schooners, the *Goodspeed* lost a fisherman, coming in to T Wharf on 8 December 1913 with her flag lashed to the fore shrouds in the traditional signal that a crewman had been lost. Captain Perry had to report that, "yesterday, while hauling his trawls in a 40-mile-an-hour gale, and a mountainous sea, Lewis Crowe, the purser, was lost."[28]

Based on records of seventy-seven schooners compiled by Gordon Thomas in *Fast & Able,* it could be expected that a vessel would lose a man every four years. At that rate, the *Goodspeed* was somewhat safer than average, losing only one in eleven years. But she herself did not surpass the fifteen year average fishing career of those same seventy-seven schooners. She was cut down and sunk in a collision with the steam trawler *Swell,* 2 June 1919, just west of Georges Bank, eleven years and one day after her launch.

South Side of T Wharf, 1:00 P.M., 9 November 1908

A STURDY, blanketed dray horse lunches as a fisherman mends a sail aboard the nearest schooner and wharf workers stroll, enjoying what appears to be a warm late-fall afternoon. About twelve schooners are in the south dock; by the end of the day, fourteen schooners will have landed 146,800 pounds of fish, mostly hake, and almost 250 fishermen with money in their pockets.[29]

Coming ashore, all of Boston spread out before the fishermen. Charlie York remembered arriving at T Wharf.

No matter what time of day or night we got in, the first thing we done was to send one of the men to pick up our mail and they was always a clerk on hand to serve us. Our girls and wives and friends had our address, Reading Room, T. Wharf, Schooner *Eva and Mildred.* For three weeks or more we hadn't heard from a soul; they might have been fog for days, with nothing to do but take your turn at grinding out a blast on the foghorn every three to five minutes. Maybe we'd been beaten by a storm, or had hard luck fishin'. Them letters was a godsend.[30]

Besides the reading room for fishermen, there was burlesque at the Old Howard Theater in Scollay Square; and Austin & Stone's Museum, with acrobatics, vaudeville, movies, and such diversions as "Elnora the picturesque tatooed lady; Solella, the harmonica king; and versatile Franks and his broom factory." Many watering holes, such as the Atlantic Cafe on Atlantic Avenue at the head of the wharf, also served the fishermen. And there were occasional amusements on the wharf, such as the "laughing haddock," a fish that made a laughing sound when squeezed, sold by a Portuguese fisherman in 1913.[31]

As was the case in every seaport, numerous ways were devised to separate seamen from their money. One writer lamented: "The wharf attracts the bummer, the beat, and the homeless. For here are rich pickings. There are men who exist on the bounty of the fisherman who, with perhaps a hundred dollars in his pocket, seeks some convenient barroom and in generous mood invites all present to partake." There were other activities, particularly for fishermen with money. Stall 19 on T Wharf served as an illicit gambling parlor, with pool and billiards, until being closed by a police raid in June 1912.[32]

Francis J. O'Hara, Jr., in the South Dock, 16 February 1911

THE *Francis J. O'Hara, Jr.,* an Italian power dory and what appears to be the schooner *Appomattox,* lie at Long Wharf, covered with a fresh dusting of

Francis J. O'Hara, Jr.

snow and surrounded by a sheet of ice. The *O'Hara* has apparently been laid up here for the last month, but will head back to sea in five days. While setting the mainsail on the way down harbor, a young fisherman from Wood's Harbor, Nova Scotia, on his first trip in an American schooner, will slip from the cabin top and drown.[33]

The *Francis J. O'Hara, Jr.*, was owned by Francis J. O'Hara & Co., "Wholesale and Commission

Dealers in All Kinds of Sea and Lake Fish," as well as fishing dories, on Atlantic Avenue. Francis Senior was an officer in the Boston Fish Market Corporation, which backed the new fish pier, and Francis Junior was an organizer of the New England Fish Exchange in 1908.[34]

The *Francis J. O'Hara, Jr.*, was built by Oxner & Story at Essex in 1904, from a design by Thomas McManus. Typical of her use, in 1910 she was fitted out for the spring mackerel season, in July turned to shacking (salting fish taken on the first baiting and icing those taken on the second baiting of each voyage) and then went winter haddocking in October. Her career out of Boston was generally uneventful, though she barely missed disaster in March 1908, when she met the *A. D. Story* bow to bow in the fog. In the brush the *O'Hara* came away with the *Story's* starboard anchor. The *O'Hara* was sold to Gloucester owners in 1912, and was ultimately destroyed by a German U-boat, after her crew had been set adrift, 21 August 1918.[35]

The *Appomattox*, built by A. D. Story in 1902, was one of seventeen schooners built on the lines of J. H. Burnham's *Boyd & Leeds* model of 1894. She fished out of Boston until being sold to Cuban owners in December 1913.[36]

Mary DeCosta in the South Dock, February 1914

IT IS A QUIET DAY at T Wharf, just before the busy Lenten season. At Long Wharf, a waterboat is almost obscured between a fourteen-dory market schooner and the three-masted, centerboard coaster *City of Augusta*. Registered in Bath, Maine, where she was built in 1884, the *City of Augusta* is perhaps delivering building materials for the custom house tower, which rises in the background.

The *Mary DeCosta* displays the white gaffs, booms, and mastheads typical of many Portuguese vessels. She was built by John Bishop at Gloucester in 1909 for the investors represented by Leandro J. Costa, Jr., a provision merchant in the North End. Her first skipper was Captain Joseph Silveira. Between 1907 and 1912, six vessels were built for the firm, each named for a female member of the Costa or Silveira family. By 1914 they also owned the famous *Jessie Costa*, formerly of Provincetown. This fleet composed the major portion of the Portuguese vessels fishing out of Boston. As in most of the Portuguese vessels, the *Mary DeCosta's* sixteen fishermen fished single dories (one man to a dory) on Georges Bank, the South Channel, and other areas frequented by the market schooners.

Mary DeCosta received a gas engine in 1918, diesel power in 1924, and continued to fish until foundering off Miami, Florida, in December 1945.[37]

One of the *Mary DeCosta's* skippers was William J. Forbes, a native of Wood's Harbor, Shelburne County, Nova Scotia. Born in 1866, he began fishing at seven and had migrated to Gloucester by the age of eighteen. Apparently he had his first command fifteen years later. His career as a skipper was particularly varied. He sailed for at least eight Gloucester and Boston firms, including John Pew & Son, Slade Groton, C. W. Brundage, and Atlantic Maritime Co. He was the first skipper of Atlantic Maritime's *Elsie*, later famous in the International Fishermen's Races, and also commanded a vessel from Bucksport, Maine. Though he was not Portuguese, Forbes commanded three vessels for the Manta family of Provincetown, and then the *Mary DeCosta* for L. J. Costa of Boston. He was still active in 1926, fifty-two years after first handling a fishing line![38]

Mary DeCosta

South Dock, T Wharf, 1:00 P.M., 4 September 1908

AT LEAST nine vessels are in the south dock: eighteen arrived at T Wharf yesterday and nine more will come in today. Two schooners have their triangular riding sails set, either to dry or to provide shade on a windless day. The near one is probably the *Stranger,* which arrived early with 16,000 pounds of haddock, 10,000 of cod, 2,000 of hake, and 500 of pollock.[39]

In the immediate foreground, alongside Long Wharf, is a swordfisherman. A man is aloft in her fore shrouds, perhaps replacing ratlines, which received a lot of use in a swordfisherman because the foretop was rigged as a lookout, with several platforms and safety lines. Safety lines identify another swordfisherman halfway down T Wharf. A waterboat is alongside.

September was nearly the end of the New England swordfish season, which began in June. However, on this day the *Valentinna,* under Captain Neally, will land 13 fish. Yesterday, Captain

South dock

Herbert Pennington brought the *Motor* in with 25 fish and the *Mabelle E. Leavitt* arrived with 21 more. The *Motor* was the highline swordfisherman of 1908, setting new records in that fishery. During five trips to Georges Bank between June and September, her crew of eight had captured 254 swordfish, worth $7,000. It was estimated that they would receive $405 apiece, while the cook, who had saved the heads, would receive $432 for his extra effort. The *Valentinna* was not far behind the *Motor;* her crewmen were expected to share $380 each for their summer's hunting.[40]

Swordfishing was a relatively recent fishery in 1908. While swordfish had been caught south of Cape Cod by 1840, one story has it that the first swordfish to arrive in Boston was sent by Charles Manta of Nantucket in 1875. It did not sell in Boston, but later, a fish found a receptive market in Worcester. The word spread and the public acquired a taste for it. By the 1880s, a sizable fleet of small schooners and sloops hunted swordfish in the summer, and many market schooners were equipped to take those they met on the banks.[41]

Power was adopted early by the swordfishermen. Commonly, a twenty-five-horsepower engine was installed to power a schooner at six or seven miles an hour. One schooner under power took twenty-one swordfish in an afternoon while nearby sailing swordfishermen lay becalmed. According to Albert Cook Church, "there are so many vessels thus equipped that it has become almost a necessity to have power aboard, and it is a very difficult matter to ship a first-class crew unless the vessel is fitted with power."[42]

The two swordfishermen here are probably the *Motor* and the *Mabelle E. Leavitt.* Despite her name, *Motor* was built without auxiliary power at Gloucester in 1903. She went swordfishing and fishing alongshore out of Gloucester until 1909. That year, she was sold to Boston owners and a twenty-four-horsepower engine was installed. She fished out of Boston for sixteen years until being sold to New York City owners in 1925. Two years later she was abandoned.

The schooner-boat *Mabelle E. Leavitt* was built at Bristol, Maine, in 1900 and fished out of York, Maine, until 1910. She probably landed her fish at Portland or Boston, depending on the time of the year and market conditions, as did the

Eva and Mildred on which Charlie York served. The *Leavitt* was sold to Boston in 1910, and five years later had a twenty-four-horsepower engine installed. In 1919 she was sold to Gloucester, and three years later went to Nova Scotia or Newfoundland.

The small market schooner *Stranger*, on the left, was a B. B. Crowninshield design, built by Oxner & Story at Essex in 1903. Designed as a fast schooner, she fished out of Boston with ten single dories. In 1910, a thirty-six-horsepower engine was added. But even with power, she was unable to find "Tug" Wilson, one of her crew who went astray in fog on the South Channel in July 1912. He rowed for three days without food or water until meeting a yacht with supplies to share. Declining to be picked up, he eventually rowed 150 miles before a towboat gave him a ride back to Boston.[43]

In 1921, the *Stranger* was bought by Captain Doughty of Bailey Island, Maine, to fish out of Portland. Five years later, she was sold to a Newport, Rhode Island, owner. He moved to Boston in 1930, bringing the *Stranger*, now with a sixty-horsepower diesel engine. She was given a ninety-five-horsepower diesel in 1935, and a year later went to New Bedford, following a long line of fishing schooners that became "Brava packets," running between New Bedford and the Cape Verde Islands. The *Stranger* disappeared from the registry after 1937.

The End of T Wharf, Winter 1911–12

IN THE DISTANCE the East Boston or Winnisimmet Ferry passes beyond idle tugs of the Commercial Tow Boat Company and the Boston Tow Boat Company at Commercial Wharf. Between them, these companies owned seventeen harbor and coastal tugs.

Two haddockers, recently arrived, lie at the end of T Wharf. The nearest one has unloaded, and the after hatch is still open. Between the dories (slightly out of focus) the open forecastle hatch, the white galley hatch, and the booby hatch,

End of T Wharf

View from the custom house tower

which most haddockers carried over the fore hatch, are visible. The sheer strakes of this schooner's dories are painted a distinctive color. Several trawl anchors are stowed between the dories and the hatch.

Spectators in winter clothing view the harbor. The man with a bundle under his arm appears to be a fisherman coming ashore. What did fishermen do in winter? A 1910 estimate placed 6,000 men aboard Massachusetts fishing vessels in summer and 1,500 to 2,000 in winter. In the winter of 1912–13, it was claimed that at least a hundred schooners were idled by the lack of fishermen.[44]

Those who continued to fish faced danger and discomfort, but might also earn lucrative shares when fish were scarce. Speaking of winter fishing, Charlie York had a heartening thought: "as long as salt water is comin' onto your hands, they won't freeze."[45] During frigid weather early in 1914, several fishermen in one schooner spent most of their earnings for ten days of fishing on medical bills to treat pneumonia.[46]

The two thirds of Massachusetts fishermen who did not fish in winter had a number of choices.

Many "downhomers" returned to Nova Scotia and Newfoundland in December, coming back to the fishing ports in the spring. The cost of living in the Provinces was about half that in the States so some fishermen could afford to relax at home for the winter. Bad weather directed others to winter jobs ashore as teamsters or longshoremen. The expanding red snapper fishery in the Gulf of Mexico offered a warm alternative to fishing in New England. In September 1909, 1,000 fishermen headed south to Florida to pursue red snapper until April, and that fishery continued to expand.[47]

View from the Custom House Tower, 1915

THE NEW custom house tower offers a bird's eye view of Boston Harbor. Between Long Wharf and East Boston, a dredge works in the channel, which runs past the vessels moored on Bird Island Flats and out around Governor's Island at the upper right. The dredging project opened up the ship channel through Broad Sound, eliminating the treacherous stretch of channel through the Narrows.

A few fishing schooners and Italian fishing boats lie at the largely deserted T Wharf, at left. A gill net is laid out on the wharf near the Italian boats. At Long Wharf, lighters lie along the north side while a United Fruit Company steamer unloads bananas on the south side. Like Long Wharf, Central Wharf maintains some shipping activity, with horse-drawn drays and a few motorized trucks coming and going. The steamer hidden at the end of Central Wharf may be the Dominion Atlantic Railroad steamer *Prince George,* running between Boston and Yarmouth, Nova Scotia. Between Central and India wharves, three big coastal steamers are berthed, two with steam up.

Although the fishing industry did not totally abandon T Wharf in 1914, it was a much quieter place thereafter. Various agencies proposed that it be used for a recreational area, a coastal trade installation, or fish processing plant, but the wharf remained unchanged, except for the Quincy Market Cold Storage Company freezer, built in 1917. On 28 March 1914, the *Boston Transcript* bade farewell to the liveliest days on T Wharf.

This is both a glad and a sorry day—the last of T wharf. That numerous confraternity which nooned itself each day amid the thicket of masts and fish carts will miss the diversion, and the whiff of romance (and fish) which it bore to many a shore-spent life. . . . A pleasant place it was; on the first mild days of spring, as now, when warm sun gilded spars and decks and crews baited up or lounged in shirtsleeves; on the golden mornings of summer when all the waters of the upper harbor were brilliant blue; on the crisp afternoons of October, wind ruffling the bay to green and white, black clouds sailing across from the southwest, and the gulls, which begin to come into the harbor only with the cold days, wheeling and squawking like a rusty block. There will be tales as wild come to the gorgeous new pier at South Boston; there will be schooners as smart and crews as doughty; but memory and affection will cling somehow to the dingy old wharf at the foot of State street.

South Boston Fish Pier, 31 March 1914

TAKEN from the new Commonwealth Pier, this view shows the new fish pier on its second day of operation. On opening day, 30 March, twenty-one vessels delivered 1,452,000 pounds of fish, and six more vessels will arrive today. The substantial, 1,200-foot wharf and the spacious slips of the new pier are a great contrast to crowded T Wharf.

The schooner on the right, with the dory sails set, is probably the *Robert & Arthur,* built by Oxner & Story at Essex in 1902. Today she is

South Boston fish pier

laden with 23,000 pounds of haddock, 8,000 of cod, and 500 of pollock. Under Captain Julius Anderson, she had the seventh best stock of 1910—$35,000.[48]

The *Robert & Arthur* had a hard existence. In March 1909, she was caught in a gale and heavy seas, which swept the dories off her deck, unshipped her rudder, and toppled her mainmast. Under jury rig, she returned to port, where she was given a shortened mainmast, making her easily identifiable. Four years later, in another March gale, she was hove down, with mastheads nearly to the water, but righted herself with only the loss of gear on deck.[49]

There was a general trend toward auxiliary power in the sailing fleet during the 'teens, and the *Robert & Arthur* was included, receiving a 120 horsepower gas engine in 1918. After twenty years of fishing out of Boston, she was sold to Halifax, Nova Scotia, owners.

North side of the new fish pier

Spacious new fish pier

A Stroll Around the New Fish Pier, 31 March 1914

THESE FIVE VIEWS were taken at lunch time. The first view looks out the north side of the fish pier. On the right is the icehouse, still enclosed in scaffolding. On the left is the schooner *W. M. Goodspeed*. Her fish were the first auctioned off as the pier opened yesterday.

The spaciousness of the 1,200-foot wharf is evident in the second view. The third view shows the south side of the pier. The paving blocks will be used to finish paving the end of the pier. The pile of blocks partially conceals the steam trawler, *Wave, Surf*, or *Crest*, which arrived yesterday.

The idle handcarts suggest that some of the wharf's seventy-five handcart haulers are still on strike. After struggling the length and breadth of the pier to distribute almost a million and a half pounds of fish the first day, this morning about half of them demanded an increase in pay to ten cents for hauling a load to the near side of the pier and twenty cents to take a cart load to a dealer on the other side. After a few hours, the dealers agreed to the increase.[50]

The final two views were taken in the pier's central alley or "center street." Here, processed fish could be loaded onto drays or trucks for shipment, with little of the congestion seen on T Wharf. At the inner end is the huge icehouse still under construction. At the outer end is the office building and auction room of the New England Fish Market Company.

South side of the new fish pier

"Center Street" looking in

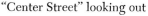

"Center Street" looking out

2 Schooner Construction

By 1900, to speak of fishing schooner construction meant, for all practical purposes, to speak of Essex, Massachusetts, an ancient settlement north of Gloucester, situated about five miles up the Essex River from Ipswich Bay. Shipbuilding dates from perhaps 1660 at Essex, and before 1850 the town became the major producer of fishing schooners. John Bishop and others built schooners in Gloucester, and a few were produced around Boston, but Essex specialized in schooner construction and monopolized the field for decades. As the

photographs show, the long-used shipyards were merely sloping plots of ground along the river bank (to allow gravity to launch the vessels) usually with a shed for storage, perhaps a mechanical saw for cutting timbers and planks, and often a few piles of lumber.

Depending on the economy, several of these unassuming yards might be in production; however, shipbuilding was sensitive to economic changes. Following the recession in 1907–08, only three vessels were launched in 1909. In better times, two yards produced thir-

Essex, low tide

ty vessels in 1901, and three yards launched thirty-seven vessels in 1902. The major yards after 1900 were those of Arthur D. Story, which produced about 475 vessels between 1875 and 1930; James & Tarr (sometimes called Tarr & James between 1910 and 1912) a consistent producer of fine-quality schooners; and Oxner & Story, which produced several innovative schooners between its establishment in 1900 and its failure in 1907.

As Dana Story pointed out in *Frame Up!* life in Essex revolved around the shipyards. The men worked in the yards: a few on specific yard crews, those with skills such as caulking or inside joinery as contract labor. Others worked in associated maritime trades, producing windlasses, spars, or ironwork. Others were employed in support services, producing food or operating stores in which credit was extended to shipwrights. And the shipwrights' wives often cooked over wood fires fueled by chips from the shipyards.

Essex was predominantly Yankee and some-what ingrown until Nova Scotia shipwrights entered the community around 1900. But despite changes in personnel, Essex remained a pocket of nineteenth-century industry well into the twentieth century.

Essex, Massachusetts, at Low and High Tide, August and October 1912

THE CAUSEWAY leads from South Essex to Essex. Along it lies the James & Tarr yard, where the *Gertrude DeCosta* is under construction. Beyond is the center of Essex and Arthur D. Story's shipyard. Upstream, beyond the causeway, is Andrews' sparyard.[1] Five miles of narrow, twisting channel, much of it navigable only at high tide by a fishing schooner of 1900, lies between Essex and the open water of Ipswich Bay.

Essex and its situation were typical of small New England shipbuilding communities of the early nineteenth century; yet its industry survived well into the twentieth century. Originally, timber

Essex, high tide

supplies, suitable shoreline, and skilled labor brought shipbuilding to a location like Essex. By the mid-nineteenth century, small yards lined the river banks, manned by second- or third-generation shipwrights, building schooners for the local Gloucester market with imported timber. By 1900, a few larger yards remained, with a combination of local and immigrant labor. Timber came from the south, from Ohio, and from Maine, but little came from Essex county. While the river was no longer ideal for producing vessels of the size built there, the skilled labor living along its banks continued to produce the wooden vesels needed in the fisheries up through the 1940s.

Leonard MacKenzie Yard, South Essex, 20 August 1912

THE SCHOONER in frame is the *Yucatan*. Her stem has recently been hoisted, shored up, and clamped in place. The cant frames, which butt against the stem as opposed to the square frames visible here resting across the keel in complete U-shaped units,

lie under the inclined plank. The light framework behind them may be a mold representing the shape of some timber. It appears that the stem rabbet, the groove in the stem which receives the ends of the planks for a tight fit, has been cut part way up. Much of the timber is probably longleaf yellow pine, since that wood resists the effects of the warm Gulf of Mexico waters, where the *Yucatan* will fish, better than the standard white oak of New England fishing schooners.[2] Three men are dimly visible, working along the *Yucatan*'s keel. On the right is the first evidence of the *Yucatan*'s sister *Arcas*.

This yard was previously used by Willard A. Burnham, and then by Oxner & Story from 1900 until their failure in 1907. Leonard MacKenzie, a native of Essex, built at several sites in Gloucester before returning to Essex to build two vessels in 1912 and one in 1913, just before his death. Owen Lantz used the yard between 1916 and 1920.[3]

The *Yucatan* and *Arcas* were two of six flush-decked schooners built for "green fishing" in the Gulf of Mexico, from a model by Cox & Stevens of New York, best known for their yacht designs. The flush deck was unpopular among fishermen as it

Leonard MacKenzie yard

James & Tarr yard

allowed boarding seas to sweep the deck, but these schooners, designed to carry auxiliary gasoline or diesel power, were considered innovative by *Yachting* magazine in 1912. Captain Clifton Smith took the *Yucatan* south after launch to fish for red snapper out of Galveston, Texas. In March 1913 she landed 54,950 pounds of red snapper, and the $99 share for each fisherman was the largest to that date. She fished under sail alone until a diesel engine was installed in 1925, and disappeared from the registry after 1932.[4]

and 139 of them in partnership with James. The James yard had built eleven of the forty-three Essex-built fishing vessels documented in Boston in 1914.[5]

The James yard was known for the fine finish given its vessels. While the Story yard contracted for a straight price on a vessel, James & Tarr operated on a "cost plus" basis. The final price of a James built schooner depended on itemized expenses and the degree of finish desired by the owner.[6]

The James & Tarr Shipyard, Essex, 13 October 1912

THE James & Tarr yard was located along the causeway between Essex and South Essex (in the distance). Here, the Boston schooner *Gertrude DeCosta* sits on the stocks, almost ready for launching. She is partially concealed by the yard sawmill. The other buildings contain the yard office and storage areas.

The second most productive shipbuilding firm at Essex was the partnership of Washington Tarr and John F. James, who built on this site from 1885 to 1912. John James and other members of the James family used this site continuously between 1838 and 1947. Washington Tarr had been active for over fifty years when he retired from the firm late in 1912. He had built more than 250 vessels,

Gertrude DeCosta in Frame at the James & Tarr Yard, 20 August 1912

SURROUNDED by a spindly set of stem supports and catwalks, the skeleton of the *Gertrude DeCosta* begins to receive its plank skin. Several of the lower strakes of planking have been put in place, and the wooden trunnel fastenings have not yet been cut off. The wide upper strake is not a plank but a ribband to hold the frames in line.

The gin pole set up at the bow can be used to skid ceiling planks up the long sloping brow and maneuver them into the hold. Perhaps the three shipwrights bending to their work are shaping a ceiling plank, as it appears wider than an outside plank. There is a pile of sawn futtocks (frame sections) by the staging at the right, where the *A. Piatt Andrew* was built in the spring. Someone

Gertrude DeCosta in frame

has left a lipped adze on the oak planks in the foreground. In the extreme background, the stem of the *Yucatan* rises between the sheds at the MacKenzie yard.

An auxiliary vessel like the *DeCosta* would cost between $25,000 and $30,000 to construct in 1912. The finely finished auxiliary schooner *Alert* was considered expensive at $29,000 in 1906, but the pole-masted, oil-engine auxiliary knockabouts *Knickerbocker* and *Bay State* of 1912 were $30,000 vessels. More like the *Gertrude DeCosta* were the knockabouts built for Boston in 1908. The *W. M. Goodspeed* and *Athena* were estimated at $25,000 apiece, with fittings, when they were constructed in 1908.[7]

Prices changed, depending on availability of timber, labor supply, number of vessels in demand, installation of auxiliary power, and other variables. However, based on the above figures, the cost of a schooner ranged from $200 to $230 per ton, complete. An otter trawler of the same period, built of steel and generally of twice the tonnage,

would cost about $290 a ton complete. The auxiliary knockabout *Evelyn M. Thompson*, sister to the *Athena*, depreciated in value by about half during her ten year existence, but a capable skipper could return several times the vessel's original cost in profits in ten years' time.[8]

Getting Out Planks at the James Yard, August or September 1914

IN THESE two views, the schooner *Robert & Richard* is being planked up alongside the causeway from Essex to South Essex. Planks sawn on the bandsaw are dragged by the yard horse to the hull where they are planed and beveled for a caulking seam before being fastened in place.

The steam-powered bandsaw did not begin to replace the hand-powered pit saw in Essex shipyards until 1884, over fifty years after it appeared in New York shipyards. Here, at the James yard, the steam engine was housed in a shed, with the saw under an overhang. Judging from the wires, the bandsaw was electrified soon after transmission lines reached Essex in 1911. The saw appears to have a quadrant arrangement allowing the table to be angled to cut on a bevel, useful since a vessel's frames, both fore and aft, had beveled faces. For maneuvering heavy timbers to the saw blade there is a derrick and a roller.[9]

On the way from the saw to the hull, the faithful yard horse plods a well-trodden path. The horse has arrived with a freshly cut plank in the lower view. Apparently it has learned to stop when one of the crew calls out and steps on the timber dragging behind.

Judging from the shavings, the nearest man is finishing planks. Near him is the handle of a broadax that he uses to bevel the edge of each

Getting out planks

Fitting a plank

with the garboard plank at the keel and have worked upward. Another common method was to begin at both the garboard plank and the sheer—uppermost—and finish at the turn of the bilge. The construction of the "stern circle," the transom planked over "tail feathers," is quite visible in this view. In the foreground is the frame for a portable windlass called a "crab," with which a few men could hoist heavy timber with relative ease.[11]

The *Robert & Richard* was launched at the end of October 1914. Built for John Chisholm of Gloucester, she was named for the sons of her captain, Robert Wharton. She arrived from her maiden trip on 16 February 1915 with 50,000 pounds of halibut, 25,000 pounds of mixed fresh fish, and 5,000 pounds of salt fish.[12]

The *Robert & Richard* only lasted about three years, being torpedoed off the Maine coast, 22 July 1918, during a German raid on the fishing fleet.

plank. Next to the ax is a jack plane used to finish the edges, forming caulking seams between planks when they are fastened in place. He is working on two thicknesses of planking: perhaps one is outside planking and the other is inner planking, or ceiling.

A schooner like the *Robert & Richard* was very strongly built. Generally, the exterior planking was white oak, perhaps 2½ inches thick below the waterline and 3 inches above. A yellow pine ceiling, varying from 2½ to 4 inches thick, was laid on the inside surface of the frames, adding great strength to the hull. An average schooner contained about 12,000 board feet of oak and 6,000 board feet of yellow pine.[10]

Fitting a Plank on the *Robert & Richard*, John F. James Yard, Essex, August or September 1914

THREE MEN of the planking crew clamp and wedge a plank into place. Apparently they began

Gertrude DeCosta Ready for Launch, 13 October 1912

THE BOW of the freshly painted vessel shines as she is readied for her launch, which will take place in two days. The shipcarver has finished her standard hawsepipe carving and block-lettered nameboard. The *Yucatan*, just visible over the sheds to the right, does not yet have her nameboard affixed. The pile of oak planks wrapped with chain will act as a drag during the launch. Attached to her bow, it will follow as the *DeCosta* takes to the water and bring her to a halt before she buries her stern in the opposite bank of the river or escapes downstream.[13]

The *DeCosta*, built for the owners represented by Leandro J. Costa, Jr., of Boston, fished with sixteen single dories to supply the Boston market with fresh fish. In 1925, L. J. Costa, Jr., moved to Gloucester and the *DeCosta* followed. That year, her 100-horsepower gas engine was replaced by a 100-horsepower diesel. The O'Hara family purchased her in 1929 and returned her to Boston. She continued fishing out of Boston until she disappeared from the registry after 1949, at the end of a thirty-seven-year career.

28

Stern View of the *Gertrude DeCosta*, Ready for Launch, 13 October 1912

THE HULL has been painted and most of the fittings are in place. Amidships, the fore boom crutch stands, with the loop of the main fiferail, not yet in place, behind it. The rails of the launching ways are visible beneath the vessel. Also visible is the aperture for the propeller, just forward of the rudder. In the foreground lies a collection of oak timber. The yard molder makes use of the natural curves of the oak to shape the double sawn frames, or ribs, of the vessels.

Howard I. Chapelle, who included plans of the *Gertrude DeCosta* in his *American Fishing Schooners*, considered her an example of the apex of New England fishing schooner design: an auxiliary knockabout with the best features of both sail and power vessels. She was fitted with a seventy-horsepower engine and, without topmasts, her sail area was reduced by one-third over similar knockabouts.[14]

After Captain Solomon Jacobs proved the value of the gasoline engine in a schooner-rigged fisherman with his *Helen Miller Gould* of 1900, the auxiliary schooner became increasingly common. It was particularly well suited in the mackerel fishery for powering through calms to get on fish or to run to market. Three auxiliaries were added to the fleet in 1901, and three more in 1902.[15]

Gertrude DeCosta ready for launch

In 1905, the first auxiliary haddocker, *Elizabeth Silsbee*, was built. While her 300-horsepower engine was too large to be profitable, she did demonstrate the benefits of an engine in the market vessels. By 1914, eleven (18 percent) of the sixty Boston schooners over fifty tons were auxiliaries, and there were twenty-eight more gas boats of various sizes. Most engines were under 100-horsepower.[16]

Numerous explosions or fires in small gasoline powered boats and in several fishing schooners, including the *Helen Miller Gould,* and the highly productive *Mary C. Santos* in 1916, illustrated the hazards of the gasoline auxiliary. Therefore, the diesel engine, developed in Germany in the 1890s, was viewed with interest.

The diesel engine relies on high compression and heat to combust crude oil and drive a piston rather than on the explosion of more volatile gasoline, making it safer, more efficient, and more economical. The pioneering work was done in Europe, and it was not until about 1910 that American manufacturers experimented with the design. While the Blanchard type, installed in the *Bay State* and *Knickerbocker*, proved disappointing, the New London Ship and Engine Company—Nlseco—obtained rights to produce the excellent design of the Machinenfabrik Augsburg–Nuremburg of Germany. Incorporated in October 1910, Nlseco was producing engines eight months later and found a highly receptive market.[17]

As demand for gasoline outstripped production, the price climbed from ten cents a gallon in 1908, to twelve cents in 1912 and nearly twenty cents in 1914. As the First World War progressed, it rose to twenty-four cents in Gloucester and twenty-six cents in Boston by 1916, with the expectation of going to thirty cents. At those rates, the initially expensive diesel engines became attractive. A 300-horsepower gas engine would burn twenty-eight gallons an hour, and at 180 horsepower, over twenty-two gallons an hour were consumed. Even a 70-horsepower engine would burn about ten gallons, or $2.00 to $3.00, an hour. Crude oil cost closer to four cents a gallon and was consumed at half the rate of gasoline, making even continuous operation feasible. A 100-horsepower diesel might burn six gallons an hour, operating for thirty cents an hour at full power. Furthermore, a diesel engine was claimed to save 40 to 50 percent in space and weight over gasoline engine installation at that time.[18]

The *Manhassett* of Boston was given a 120-horsepower Nlseco diesel engine in the summer of 1914, and proved the value of the diesel auxiliary to the New England fishing fleet. She was reported to make nine knots easily under power. Within a decade, the diesel or crude oil engine became the standard auxiliary power for fishing schooners and the power plant of choice for the rapidly growing fleet of small and medium size otter trawlers.[19]

Stern of the *Gertrude DeCosta*

A. D. Story yard

Arthur D. Story Shipyard, Essex, 13 October 1912

FOUR VESSELS are on the stocks: a schooner built on speculation that will be the *Delphine Cabral* of Provincetown; the gasoline-powered fishing boat *Mary F. Ruth* of Gloucester; the *Knickerbocker,* built to fish on the West Coast; and the *Ruth* of Boston.

Shipbuilding began on this site about 1660 and continues to this day. Arthur D. Story, one of a long line of Essex shipbuilders (though not a trained ship carpenter himself) went into a ship-building partnership at age eighteen in 1872. In 1880 he went into business on his own, gradually consolidating several yards into one on this site. Between 1880 and 1932 Story built 406 vessels, most of them here. Shortly after 1908 he also became involved in the Boston fish wholesaling business as an investor in the Story-Simmons Company.[20]

Almost half of the Essex-built fishing vessels enrolled at Boston in 1914 were built by A. D. Story. His yard had a reputation for solid vessels without frills for a reasonable price.[21]

Three stages of construction

Three Stages of Construction at the A. D. Story Yard, August or September 1914

ON THE RIGHT, the *Henrietta* is visible as a keel and three frames. Framing has begun aft, where the deadwood will be placed. The framing stage is set up along the keel to continue the assembling and setting up of frames forward toward the bow.

In the middle, the *Somerville* is being planked. It appears that the planking crew has begun at both the garboard (bottom-most plank) and sheer plank (top plank). They have not yet reached the turn of the bilge, where the shutter plank will finish the job.

On the left, the *Reading* is nearing completion. One man puts finishing touches on the trunk cabin aft while another works near the windlass at the bow. The *Reading* has a light-colored priming coat on her topsides.

The *Reading* was launched in mid-September 1914, for Captain Hickey. She was used as a market schooner out of Boston, primarily for haddock and hake. While it was reported shortly after her first trip that power would be installed in the

Reading, she was still listed as a sailing vessel in 1919. Apparently the next year she was sold to Grand Bank, Newfoundland, owners, and was fishing with a St. John's registry in 1925.[25]

The *Somerville* was built for Captain Felix Hogan, a native of Newfoundland, who had been a skipper for the Atlantic Maritime Company since about 1900. He was master of the *Elk* until *Somerville* was launched. She was named for Hogan's adopted town. Hogan used the *Somerville* haddocking and halibuting out of Boston until 1918, when her port of enrollment was changed to Gloucester. In 1920, Hogan made plans to build the *L. A. Dunton* and sold the *Somerville* to Newfoundland.[23]

The *Henrietta* was launched in February 1915, about six months after this photograph was taken. Her lines were taken from the schooner *Angeline C. Nunan,* built by Story the previous year. With a crew of fifteen, she fished out of Boston under sail until a sixty-horsepower gasoline engine was installed in 1923. She disappeared from the registry in 1937.[24]

Story Yard, 13 October 1912

HERE, the Story yard is seen from another angle. Again, the vessels are the *Delphine Cabral,* power-boat *Mary F. Ruth, Knickerbocker,* and *Ruth.* Essex vessels were commonly launched while laid over on one bilge, thereby entering the shallow, narrow Essex River with less than their full depth. The heavy planks placed under the *Knickerbocker*'s starboard bilge suggest that she may be launched that way when she goes overboard tomorrow.

The man in the boat is Peter Hubbard, an outside joiner who often worked for A. D. Story. He has done his share of planing on the schooners in the background. Like a large percentage of the Essex shipwrights after 1900, he was from the Maritime Provinces. With the development of iron and steel shipbuilding and improvements to the steam engine in Great Britain, wooden shipbuilding declined in New Brunswick and Nova Scotia. The skilled ship carpenters took up their tools and sought their familiar trade in surviving American yards from Bath, Maine, to Essex, to Noank, Connecticut. Many of them were French-Canadian, adding a new dimension to the small, tightly bound Yankee communities.[25]

Outside Joiners Work on the *Knickerbocker,* A. D. Story Yard, 20 August 1912

FOUR VISITORS investigate the construction of the *Knickerbocker.* On the staging, two men

Story yard

Outside joiners

smooth the outside planking; at least three others work on deck. The joiner on the left has been planing the topsides while his companion eyes the planks aft. White bungs cover the heads of the iron spikes that hold the butt ends of each plank. The numbered stanchions, which will support the rail, are set down between the frames and trunnelled through the planking and ceiling. Back aft can be seen the propeller aperture forward of the sternpost, which will be filled in with a removable block. This provision for power was becoming standard by 1910, allowing an auxiliary engine to be added on the centerline without extensive rebuilding of the vessel. With two engines, *Knickerbocker* has a bearing for each shaft extending through the planking, one of which is visible to the right of the man in the white hat.[26]

The *Knickerbocker* was built because a very rich fishing ground for halibut and cod lies off the coast of Alaska, and New England fish deal-

ers were quick to exploit it for their markets. Initially, firms such as the New England Fish Company invested in western fleets to develop them. The earliest vessels were brought around from New England or copied from New England models, but the sheltered inside passage from Seattle to Alaska was not suited for sailing vessels. Quickly, Northwest fishermen adopted power vessels, the ultimate examples being the steamers *New England* and *Kingfisher* sent west by the New England Fish Company.

In 1911, the New England dealers decided to cash in more directly on the Northwest halibut bounty. Three modern auxiliary knockabouts, the *Athena, Alice,* and *Victor & Ethan,* were sent 'round Cape Horn amid much fanfare. The New England Fish Company asked Thomas McManus to design what they hoped would be the most modern and efficient halibut vessel on the West Coast—an oil-engine auxiliary knockabout. Two schooners were built to this model, the *Knickerbocker* at Essex and *Bay State* at Gloucester. Each cost $30,000.

In effect, they were designed as power craft with auxiliary sail. Power was provided by two 100-horsepower Blanchard oil engines designed to burn a very cheap, low-grade oil. Because of these economical engines (turning 3-blade feathering propellors, powering them at 10½ miles-per-hour) their schooner rigs carried a sail area of 4,500 square feet, half the usual size. They carried 7,000-gallon fuel tanks, enough for 12 days of continuous motoring. *Knickerbocker* had berths for 24 in the forecastle and 4 aft.[27]

The *Knickerbocker* departed for Seattle in March 1913, and took over 150 days to make the passage. This long passage was only the first disappointment in her career. While advanced by New England standards, her exposed wheel and lack of both power winches and water pumps for cleaning fish made her inferior to the modern halibut vessels in the Northwest. If the experience of the *Bay State* was indicative, the *Knickerbocker*'s Blanchard engines were temperamental and difficult to start, further reducing her efficiency. The *Bay State* never made the western passage. *Knickerbocker* fished for a time, drifted into unspecified service, and disappeared from the registry after 1919.[28]

Pacific fishing was also disappointing to New England dory fishermen seeking alternative employment. Reports of high prices paid for halibut, and lucrative offers extended to New England men by Pacific vessel owners lured fifty men west in 1913, with another hundred planning to follow. They soon learned that they had been called in as strikebreakers in an area that was a hotbed of labor organization at the time. They also found that conditions warranted a fishermen's union. Fishermen were exploited by the railroads going west, and by boardinghouses and outfitters. Passage to and from the fishing grounds was weeks long, and living conditions aboard the Pacific halibut vessels were primitive compared to the accommodations in New England vessels. Nor was there the incentive offered by the share system of payment, which prevailed in New England.

The respected Gloucester halibuter, Captain Lemuel Spinney of the *John H. Hammond*, was sent west to investigate, and fishermen who wished to return to Massachusetts were assured they would find sites aboard New England vessels.[29]

Like the New England fishermen, the New England schooners, except for the *Knickerbocker*, returned to their home ports after a short while. The *Athena*, *Victor & Ethan*, and *Alice* had all returned to Massachusetts by the end of 1915.

Bow View of the *Delphine Cabral* on the Stocks, A. D. Story Yard, 13 October 1912

SHOWING OFF her fine underwater shape, the vessel sits on the stocks. Some of her gear, including the wheel, lie to the left, in the shadow. Next to her, the powerboat *Mary F. Ruth* is almost planked up. Beyond is the stern of the *Knickerbocker*.

Delphine Cabral

The *Delphine Cabral* was one of at least eleven schooners Arthur D. Story built on speculation between 1899 and 1915. While the yard turned out seventy-two vessels in the prosperous years between 1900 and 1905, eighteen in 1901 alone, orders sagged after 1905 to an average of five a year. To keep his shipwrights at work, especially in winter, Story built at least one vessel on speculation almost every year between 1906 and 1915. Two of the three vessels he built in the recession year of 1909 were begun on speculation. Generally, it was not difficult to dispose of the vessels, but their construction required Story to risk his own capital.[30]

This vessel was built through the summer of 1912. Before her launch in December, she was bought by Joseph Cabral of Provincetown and christened *Delphine Cabral*. A finely modeled vessel, she attracted much attention when she brought her first trip of fish in to T Wharf in January 1913. Captain Joseph Enos used her fresh fishing for the Boston market on Georges Bank, the South Channel, and other local waters, like most Provincetown vessels of the period. The *Delphine Cabral* was sold to owners on the island of Martinique in 1919.[31]

Decks of *Knickerbocker* and *Ruth* at the Story Yard, 13 October 1912

ON THE LEFT, the clean, fresh deck of the *Knickerbocker*, prepared for launch tomorrow, shows off the efficient design of a fisherman's deck to advantage. From the bow are the iron for the forestay on the stemhead, the furling plank with forecastle skylight beneath, gallows with jumbo boom fitting and pawl post, and the windlass just forward of the forecastle hatch. Between the hatch and the mast partners and galley hatch is the jumbo sheet horse, an iron to hold the block of the forestaysail sheet. When the sheet horse was placed in this position, the fishermen sometimes faced a deadly, whipping block as they emerged from the forecastle.[32]

The smaller *Ruth* has only one hatch into the fishhold, aft of the break in the deck (the forward aperture will be the forecastle hatch). Her windlass, made locally, shows well.

The carefully constructed after cabin was almost certainly produced by Edwin Perkins, one of Essex's finest woodworkers. Through the most productive years of the Essex yards, during which the finest vessels joined the fishing fleet, Perkins was employed building the after cabins. He was still working in his eighties.[33] This cabin is a beautiful example of his work.

Several tools have been left out, including a crosscut saw with a trunnel for a handle, an auger, a spike set, and several clamps.

The Boston schooner *Ruth* was used as a market fisherman. Despite her relatively small size, she was given a 236-horsepower gasoline engine at the end of 1918. In July 1919 the tug *Piedmont* rammed and sank the *Ruth* as she got under way from the South Boston Fish Pier. She was seven years old at the time of the collision but, because of her large, new engine, she was valued at $20,000. The Boston Insurance Company, which seems to have insured many of the Boston fishing schooners, held a policy for three quarters of her claimed value. Raised by the T. A. Scott Wrecking Company, she was repaired and returned to service.[34]

Late in 1923 she was sold to the Pensacola, Florida, red snapper fleet and her 236-horsepower engine was replaced by one of 72-horsepower. She stranded on Alacran Reef, off Mexico, with no loss of life, 1 January 1925.

Knockabout Ketch Yacht *Finback* at the Atlantic Maritime Co. Wharf, Gloucester, May 1916

ALTHOUGH there was a spar yard in Essex, most of the large schooners were towed around Cape Ann to be rigged in Gloucester. The job of rigging, ballasting, and outfitting usually required up to a month. However, in one notable instance, the schooner *Monitor* began her maiden trip only eleven days after launch. The riggers at the Burnham shears needed only two hours to step her lower masts and set up the standing rigging![35]

Finback was built by A. D. Story on the lines of

Knickerbocker and *Ruth*

Finback

Following the Civil War, the great increase in yachting provided opportunities for consistent employment of naval architects. Beginning in the 1880s, yacht models increasingly influenced fishing schooner designs. Men such as Denison J. Lawlor, Edward Burgess, and Bowdoin B. Crowninshield began as yacht designers and, when approached for fishing schooner designs, applied some elements of contemporary yachts. The fine-lined, long-ended, heavily sparred fishing schooners common after 1900 owed much to these designers. According to *Rudder*,

when the late Edward Burgess began to acquire his reputation for fast schooners he was soon commissioned to design fishing schooners, and such vessels as the *Carrie Phillips, Fredonia*, and others, added greatly to his fame. Since then yacht designers have turned out many of the fishing fleet, and the success of the Crowninshield boats has brought this designer a considerable amount of this class of work. Crowninshield took bolder steps in applying the features of yacht design to his fishing schooners than had any of his predecessors, and this gave speed, while his knowledge of the conditions which the fishermen meet on the Banks in summer and winter enabled him to meet all other requirements.[37]

Also, the knockabout bow, without bowsprit, demonstrated by the *Finback*, was a feature of small racing craft developed in the 1890s and applied successfully to fishing schooner design just after 1900.

The effects of fishing schooner design on yachts were more limited but did occur. The famous *Fredonia* was used as a yacht for a year before going fishing, and the *Finback* was an example of a very powerful fishing schooner hull adapted for yachting. When *Rudder* illustrated Crowninshield's design for a sixty-foot red snapper fisherman, it noted, "this type of boat at once appeals to yachtsmen who want good sea-going cruisers, and Mr. Crowninshield has prepared plans . . . showing the vessel converted to a yacht."[38] Perhaps the finest yachting tribute to the fishing schooner was the series of *Malabar* schooner-yachts designed by John Alden through the 1920s.

the McManus-designed fishing schooner *Catherine*, which was a modification of the *Bay State/Knickerbocker* model. The *Catherine*, built in 1915, was the largest and fastest of the knockabouts. The *Finback* was built for C. H. W. Foster of Marblehead. After two years as a yacht and a year as a freighter, she was rerigged as a schooner to go whaling. She was lost near Cape Fullerton, Hudson Bay, 23 August 1919, on the last American whaling voyage to Hudson Bay.[36]

3 Fitting Out

FITTING OUT refers both to periodic maintenance of vessels and to the frequent resupplying of the vessels as they delivered fish and prepared to return to the fishing grounds. The first form included hauling of the schooners for painting and bottom work on marine railways in East Boston, Chelsea, or Gloucester. Sails needed repairs or replacement, giving work to sailmakers in Boston or Gloucester. Riggers and sparmakers in both cities also found frequent work among the weather-beaten fishing schooners.

Fitting out the vessels with supplies and fishing gear was a daily activity at T Wharf, but the seasonal aspect of the fisheries lent a definite seasonal variety to the process. Year round, large and small "market" schooners visited the ledges of Massachusetts Bay, the Gulf of Maine, Middle Bank, Georges Bank, and the South Channel east of Nantucket, making quick trips to bring back whatever fresh groundfish were in season. Almost daily, they could be seen taking on food, water, bait, ice, and other supplies. Often, their fishermen could be seen overhauling and baiting their trawls in preparation for their first day of fishing.

In winter, from about November through March, the haddockers loaded ice and fresh provisions for their one- or two-week trips to Georges Bank, the South Channel, and other prime haddock grounds. A few schooners might be seen loading provisions, but no fishing gear, to make the run to the south and west coasts of Newfoundland, where American fishermen had rights to bring back frozen herring caught by the local fishermen. Small schooners from Eastport and Lubec, Maine, might also be seen shipping supplies after discharging their cargos of herring and sardines.

With the approach of spring, some haddockers left their tubs of trawl and haddock dories ashore and took on the large halibut dories and canvas "skates" of heavy halibut trawl. They loaded more ice and provisions than did haddockers, for their four- to nine-week trips to the distant banks. A few Boston schooners went to Gloucester in mid-March to fit out for mackerel seining. They took aboard one or more huge purse seines, one or two seine boats, which had been stored and overhauled during the winter, and the other incidental gear of the mackerel fishery. After a trip south to meet the mackerel somewhere off Virginia, these vessels worked their way north, landing their catches and resupplying in Boston by late June. A few vessels took aboard mackerel gill nets and worked in the vicinity of the mackerel seiners.

Late in the spring, small vessels were withdrawn from the shore fisheries and fitted with bowsprit pulpits, safety lines aloft, a few dories, and harpoons for swordfishing. Some of the market schooners also took aboard swordfish gear at this time to capture the occasional swordfish they met on the banks.

In September or October, the swordfish disappeared, the mackerel began to run out, and the halibut became even harder to find. At this time, many vessels were overhauled

John Hays Hammond

and fitted out anew for the haddock fishery or market fishing for the winter season. Others were laid up to await spring.

Throughout this seasonal cycle, the provision merchants, the ice and coal dealers, purveyors of fishing gear and fishermen's clothing, chandlers, restaurant and tavern keepers, and boardinghouse keepers became part of this ongoing process.

John Hays Hammond on the Parkhurst Railway, Gloucester, 13 October 1912

THE *Hammond* and a small mackerel seiner sit on the busy Parkhurst Marine Railway on the Gloucester waterfront. A staging has been set up

around the *Hammond,* perhaps for caulkers, and certainly for painters. The *Hammond*'s sails have been unbent, probably for repair in one of the local sail lofts. The mackerel seiner, identifiable by the seine roller on her port rail, illustrates the blocking used to support a schooner out of water.

Only five days ago Gloucester's crack halibuter, Captain Lemuel Spinney, brought the *Hammond* in from the Gully, just northeast of Sable Island, with 15,000 pounds of halibut and 70,000 pounds of mixed fish for a $2,800 stock.[1] The *Hammond* shows the wear of five or six months of halibuting. After a brief overhaul, she will enter the haddock fishery while Captain Spinney spends a well-deserved winter at home.

John Hays Hammond, wealthy mining engineer, was a prominent citizen of Gloucester who thought enough of his adopted town to donate a Fishermen's Home on Eastern Avenue.[2] His namesake was launched by James & Tarr at Essex on 25 June 1907. Designed by Thomas McManus, she was known as a fast vessel, and under Captain Spinney had a very successful halibuting career, several times being highliner of the Gloucester fleet. She was the first halibut schooner to carry dories stacked upright on deck, and the first halibuter to have a gasoline deck hoister (installed in 1915).

Six months before this photograph was taken, the *Hammond* was involved in a tragic, but not unique, accident. On 16 April 1912, while maneuvering in a dense fog near Sable Island, the *Hammond* struck the Nova Scotia salt banker *Uranus,* sending her to the bottom. Most of her crew escaped in dories. Seven men and boys were able to board the *Hammond,* which returned to Gloucester with a damaged bow and shattered bowsprit. She was back at sea again in July. Sold to Newfoundland at the end of 1916, she was sunk by a German U-boat on 27 July 1917.[3]

Each year, fishing schooners were hauled at least once, or perhaps twice, between fishing seasons. Gloucester marine railways specialized in fishing schooner work and were apparently a bit cheaper than the East Boston railways. Richard Green of Chelsea, who built several schooners, seems to have run the principal repair yard in the Boston area.[4]

Commonwealth at the South Boston Fish Pier, about 1922

IT AS a slow day at the fish pier, with only a few schooners and a sloop-boat in port. A waterboat or gasboat is tied off at the *Commonwealth*'s bow.

Apparently, the *Commonwealth* is being overhauled. Her dories, numbered and stencilled with her name, are on the wharf. Like the knockabout schooner next to her, she has only her jib and jumbo (forestaysail) bent on, and the fresh white of her mastheads, gaffs and booms, and deck furniture suggests the painters have recently finished.

The *Commonwealth* is a good example of the large auxiliary knockabout design that Chapelle considered the high point of New England fishing schooner development.[5] She was built for the owners represented by George F. Grueby of Boston, whose syndicate in 1914 managed four schooners, all knockabouts. The *Commonwealth* was modeled after Thomas McManus's *Frances S. Grueby* design, built for Grueby in 1912.

The James Yard launched the *Commonwealth*

1 October 1913. She had two fifty-horsepower gasoline engines for auxiliary power, and also carried a power hoister on deck for hoisting fish from the hold and to help with setting sail and weighing anchor. Captain Frank Watts took command and used her for single dory haddocking with sixteen dories, though she had accommodations for twenty fishermen, with fifteen bunks in the forecastle and five in the cabin. Shortly after her launch, the *Gloucester Daily Times* noted, "Several handsome pictures and a large glass mirror from friends and organizations have been hung in the cabin as evidence of the high regard and popularity in which the skipper is held." Under Captain Watts, the *Commonwealth* proved to be a consistent performer, in 1914 stocking $44,000, fifth best among the market vessels.[6]

Ironically, the *Commonwealth* sank her sister, the *Frances S. Grueby*, in a collision in Boston Harbor in 1921. She herself caught fire and burned on Brown's Bank, 8 April 1927, with the loss of twelve of her twenty fishermen.[7]

Commonwealth

Bending on Sail Aboard the *Aspinet*, T Wharf, about 17 February 1914

THE *Aspinet* has few companions in T Wharf's south basin as her crew bends on a new mainsail. The *Aspinet* was the only vessel to arrive on 17 February, delivering 12,000 pounds of haddock, 18,000 of cod, and 1,000 of hake, but losing twenty tubs of trawl on the gale-ridden fishing grounds. The ice caking the port fore shrouds as high as the lightboard suggests the kind of weather she faced on the way in. January 1914 brought the coldest weather in seven years. Consequently, the fishing was very hard, but vessels landing fish profited nicely. When Captain Brigham brought the *Aspinet* in from two weeks on Nova Scotia's Cape Shore with 85,000 pounds of haddock (100,000 of fish altogether) the $4,400 gross stock (yield) was the largest for a haddock trip to that time. The fishermen each received $135, a very good share.[8]

The normal conditions of fishing were generally hard on the vessels, and particularly on their rigs. For instance, in April 1913, while tending her dories on good fishing ground, the knockabout *Pontiac* jibed, snapping her main boom and tearing her mainsail. She limped home to T Wharf to refit, losing a week of fishing by the time she returned to the banks.[9] This was minor damage compared to the numerous cases of dismastings and vessels damaged in storms.

Records of Gordon & Hutchins, sailmakers of Commercial Street, Boston, suggest the amount of work done on fishing schooner sails. In 1915, the *Aspinet*'s sister, the *John J. Fallon*, had her sails overhauled spring and fall. Each of her lower sails was repaired at least once, and a new flying jib (ballooner) and main topsail were produced. The cost of all of the work was less than $155. At this rate, after the initial investment of approximately $1,500, to produce a schooner's sails (which would last about three years), maintenance costs for sails added little to the overall maintenance of the vessel.[10]

The sailmakers who kept the *Fallon*'s sails in order made forty-five cents an hour in 1915. On an hourly basis, this was probably a bit better than

Bending on sail

the fishermen's earnings, but changes were taking place in the sailmaking business. Only since 1908 had local sailmakers begun to use sewing machines. With mechanization and changing demands on sailmakers, they became more like other urban industrial operatives than craftsmen, and they too began to organize. Gloucester sailmakers formed a union after a strike of shoreside workers in 1902. The sailmakers employed by Boston's ten firms were members of the Sailmakers' Protective Association. Sailmakers in two Gloucester lofts went on strike for three weeks in 1910, and in September 1913 they struck again, forcing work to be sent to sailmakers in the less developed outports of Maine and Nova Scotia.[11]

The iceman

The Iceman, T Wharf, Winter 1912–13

AN ICE DEALER unloads his wagons as a fisherman hands ashore the tackle for lowering the blocks of ice into the hold. The cart on the right contains a can of kerosene for the lights aboard. Note the dory sails lashed to the main rigging.

Ice was a particular necessity for the market fishermen, which brought their fish back fresh. Depending on the season, they carried varying amounts of ice. Haddockers might carry twenty tons in warm weather, but less than ten tons in mid-winter.[12] Halibut schooners, fishing in warmer weather and making longer trips, might carry twenty-five to fifty tons.[13] Swordfishing vessels required ice, and mackerelers often carried ice as well.

Natural ice was preferred to artificial since it was not as cold and, therefore, kept the fish chilled rather than freezing them.[14] The ice was cut on ponds in eastern Massachusetts and stored in icehouses in the port towns. The Union Ice Company, partially financed by T Wharf fish dealers, had a virtual monopoly on the T Wharf ice supply from 1884 to 1914, supplying thousands of tons of ice each year. Not until November 1913 was the Union Ice Company unable to meet the fishing fleet's demand for ice. By that month, supplies had dwindled so low that ice had to be brought from outside Massachusetts for the first time in thirty years.[15]

A Provisioner on the Wharf, about Fall 1913

A PROVISIONER and his bedraggled white horse pose on this wet afternoon on T Wharf. Probably a combination of rain and a strong southeast wind has caused flooding in Boston Harbor, and T Wharf itself is partially inundated. A large fleet, including several of the steam trawlers, is in port. The provision cart has a variety of boxes and kegs of foodstuffs to resupply a vessel's larder.

One Boston provision merchant who did a lucrative business supplying the fishing fleet was Leandro J. Costa, Jr., of the North End. Related directly and by marriage to several Provincetown fishing families, Costa had good connections with the fishing industry. By 1914, he was principal owner or agent of five Boston fishing schooners, thus accruing a profit from some of the vessels he supplied. John J. Fallon, a Boston grocer, also invested in schooners, including the *John J. Fallon*. It worked in reverse as well. H. Dexter Malone, schooner captain and first skipper of the otter trawler *Spray*, was managing owner of several Boston schooners, and by the early 'teens was dealing in provisions as well.[16]

Who paid for the food? Traditionally, aboard the haddockers and market schooners out of Bos-

43

A provisioner

most fish during the trip. Finally, the remainder was divided equally among the crew, including captain and cook. In other trawl fisheries and the mackerel seiners, the owners usually supplied the provisions and gear, and received half the stock. During the years before the First World War, fishermen expected to receive a $20 share on every $1,000 of the gross stock.[17]

A Cold Day at T Wharf, 1911

TWO SCHOONERS, recently arrived, are being resupplied for their return to the fishing grounds. The foredecks of both schooners are covered with a combination of salt spray ice and fresh snow. The outboard schooner, a knockabout, still has her rigidly frozen jumbo and jib set. The near schooner has salt ice high above her lightboard, evidence of a bitter trip in. The dories show signs of hard usage. The near one has a broken seat riser. Note the leather on the end of the starboard riser, to keep the trawl line from snagging. Ahead, two fishermen aboard the *Washakie* apparently shake the reef out of the icy mainsail before furling it.

Between the two schooners stands a waterboat man, writing up a bill for filling the water tanks of one of the schooners. Gannon and Plunkett supplied water to most of the fishing schooners at T Wharf. The larger schooners had 1,000-gallon tanks under the forecastle sole. In 1909, at half a cent a gallon, it cost $5.00 to fill those tanks.[18]

Supplies are being brought aboard the near schooner. They include a crate of Florida oranges. The quality and amount of food carried aboard a fishing schooner was prodigious. It was generally admitted that fishermen ate better than other mariners, and in many cases better at sea than they could afford to eat at home. The *Fishing Gazette* asked:

How much sugar do you suppose twenty-three men will consume in one week aboard a fishing schooner? About 150 pounds. How much rolled oats and oatmeal? A bushel basket full of two pound packages. How much meat? Half a steer for roasts, steaks, stews, and corned beef. Potatoes? About three bushels. Pickles, bacon, and cabbage find a place aboard the vessel in large quantities. Then there's the salt pork

ton, the owners provided only the vessel, sails, and anchors, and received a quarter or a fifth of the net stock. A common approach was to take the gross stock and first subtract the "great generals," which were the costs of bait, ice, wood, coal, towage, wharfage, and fishing lines. Then, "skippership"—2½ percent for the skipper—was deducted, and 25 percent was set aside as the vessel's part, used to cover maintenance on the vessel and any profits to be paid to the owners. "Small generals," which covered the costs of provisions, were then set aside, and a highline bonus of $2.00 to $5.00 was given to the men in the dory that caught the

44

for fish chowders and fried fish, for crews of fishing craft are notoriously fond of sea food.

The vessels are kept stocked with provisions to last a week or ten days longer than the time generally required to make a trip to the grounds and return.[19]

Another report adds hams, sauerkraut, beans, rice, turnips, cheese, eggs, molasses, and the indispensable supplies of coffee and tea.[20]

In the haddockers and market vessels, fishing on quarters or fifths (the owners supplied only the vessel and took a quarter or fifth of the return), the crew paid for the food. In the seiners and other vessels fishing on halves, the owners provided the vessel and outfit, including food, and took half the return. Between 1910 and 1914, a four-week trip required $250 to $300 worth of food.[21]

Supplies Coming Aboard, T Wharf, Winter 1912–13

A WINTER HADDOCKER fits out. A wagonload of coal has been delivered in bags for the galley range forward and the after cabin stove. Most of it will be stored in the coal locker in the forecastle. There may be coal storage under the cabin sole (floor) but generally the gang in the cabin took turns filling the scuttle from the forecastle supply. Both stoves burned constantly; yet, considering the few fires caused by stoves, they were less hazardous than gasoline engines.

The four fishermen in oilskins load ice into the after hold. One on the wharf attaches the tongs, two on the gurry box haul on the tackle to hoist

Supplies coming aboard

"Charlie Noble," "Joe LeCost," and the cook

"Charlie Noble," "Joe LeCost," and the Cook, Onboard the *Aspinet*, T Wharf, about 17 February 1914

THREE essential characters aboard a fishing schooner in winter are visible in this view of the *Aspinet*. While the crew bends on the new mainsail, the cook takes the opportunity to stock up on provisions. Here, a man who may be the cook glances up the wharf as he carries kerosene and other supplies to the forecastle.

If anyone was second officer on a fishing schooner, it was the cook, or "cookie" as fishermen often called him. While he did not take command in case of accident to the captain, he had responsibilities second only to the skipper. He managed the larder, keeping the fishermen well fed and happy for the duration of the trip. He was up before the crew, at three or four in the morning, to prepare three or more meals for twenty men each day, and to keep the "shack locker" full of pies, cookies, and left-overs for periodic "mug-ups" by the crew. He kept the running lights and other lamps filled with kerosene and burning properly. He helped tend the last dory to leave and the first to arrive back at the schooner during the daily fishing, and helped the skipper sail the schooner in the meantime. A mackerel seiner's cook often sailed the vessel alone while the crew was in the seine boat, perhaps peeling potatoes as he sat at the wheel.

For his labors, the cook commonly received a share equal to each fisherman's, plus a wage, sometimes making him the best paid man aboard. A good cook often remained with a good captain. In 1909, the *Tattler* had a rare cook. Not only were his meals good, he made a lunch box for each dory as well.[23]

Behind the cook, "Charlie Noble" is barely visible. Traditionally, Charlie Noble was the cap of the galley stove pipe. The *Aspinet* has a long

the blocks aboard, and one fisherman on deck directs the blocks through the hatch to be placed in the fish pens. Ice and bait stored aft keep the vessel in proper trim.[22]

horizontal pipe, one of several designs devised to keep a draft while discouraging water from finding its way below.

The piece of ragged canvas lashed to the fore shrouds is "Joe LeCost."[24] The coating of ice as high as the lightboard suggests how uncomfortable the half-hour to two-hour-long night watch on deck could be for winter fishermen. Dorymates usually stood watch together, each spending half the watch at the wheel and half on lookout. Joe LeCost provided some shelter as the lookout stood to windward, watching for the lights of other shipping. The torn canvas suggests that the *Aspinet* spent some time with her starboard rail under water on the way home.

Bay State at the South Boston Fish Pier, 31 March 1914

LUNCH HOUR SPECTATORS enjoy the spring sunshine at the end of the new wharf. Suddenly quiet, T Wharf lies in the distance, just to the right of the *Bay State*, near the rising skeleton of the custom house tower.

The *Bay State* has just landed the first trip of halibut at the new fish pier. Her dories are stowed upside down in the traditional manner of halibut schooners. A waterboat is alongside, refilling her tanks while provisions come aboard. Note the chafe marks worn into the starboard bow by the heavy anchor chain.

The *Bay State* was the sister of the *Knickerbocker*, seen earlier at Essex. She was built at Gloucester by Owen Lantz for the New England Fish Company, but never made the trip to the Pacific as intended. Rather, she was used for halibuting, haddocking, and shacking, with a Portland, Maine, documentation.

Frederick William Wallace remembered the *Bay State* arriving at Canso, Nova Scotia, on her maiden voyage in 1913. She had already escaped destruction once, getting dangerously close to Cape Sable because her cast-iron exhaust pipe threw the compass off by several degrees. The *Bay State* was a handsome vessel: Wallace likened her forecastle to a Pullman sleeper, with varnished woodwork and green curtains on brass rods giving privacy to each berth. Her Blanchard crude-oil engine attracted interest among captains who were considering auxiliary power. Captain Norman Ross of the *Bay State* invited Wallace and his skipper, John Apt of Digby, to watch as the *Bay State* got under way.

The engineer had the torches blowing on the Cylinderheads to preheat them before starting—necessary in crude-oil engines of that vintage. Advised by the engineer that he would be ready to start in a minute, the Skipper was at the wheel ready to go.

'Are you ready!' he called below.

'All ready!' cried the engineer.

'Heave up!' shouted the Skipper to the men at the windlass for'ad. The anchor was broken out and the order was given to start the engine. There followed a series of muffled explosions. Then came a halt. A few more coughs from the motor, and silence. The big schooner was swinging in the tide and wind and the Skipper was anxiously glancing at the vessels anchored all around him in the restricted harbour.

He hailed the engineer again just as that individual came running up from below with an oil can. 'Where the hell are you going?' came an excited query from the Skipper.

'I got to get some gasoline for the air-pump, Cap'en,' explained the other. 'She won't start—'

'But we're underway, man—the anchor's up—'

'Sorry Cap'en, but you'd better let it go again—'

With a hurried glance at an anchored schooner towards which the *Bay State* was drifting, the Skipper lost his temper and forgetting our presence and the lecture he had given on the virtues of auxiliary power, he exploded in a roar that could be heard all over Canso. 'Dam' and blast you and your jeesly blank blank engine! I'll take her out under sail!' And to the gang he bellowed; 'Away ye go on y'r fores'l and jumbo! I'd sooner trust my canvas than any bloody useless engine!'[25]

Although the *Bay State* was built as a polemasted auxiliary, these problems with her engines, and the open water conditions of the Atlantic, caused her to be rerigged with topmasts. One is missing in this view, but she has just been driven in from the Grand Bank, perhaps losing it on the way. Departing the Grand Bank at 8 A.M. Saturday, she came alongside the new fish pier at 11 P.M. Monday (last night), averaging over eleven knots for the 800 mile trip. The 40,000 pounds of halibut she brought in were worth $2,500 to $3,000, and

Captain Ross received an extra $100 as a prize for landing the first full trip of halibut.[26]

Captain Archie MacLeod took over the *Bay State* in August 1914, keeping her until his schooner *Catherine*, a modification of the *Bay State* design, was completed in October 1915. Under many notable skippers, and with a more reliable oil engine after 1917, the *Bay State* continued a successful career until stranding at Liverpool, Nova Scotia, in December 1927.[27]

Ethel B. Penny in the North Dock, T Wharf, 2 May 1913

JUDGING from the suits and vests visible here, the *Penny*'s crew has just come aboard after a few days in port. They arrived on 30 April with 6,000 pounds of haddock, 4,000 of cod, and 5,000 of hake, only to find a large fleet at the wharf and low prices prevailing.

Up forward, several fishermen ponder the *Mary*'s long main boom protruding between the *Penny*'s fore and main rigging. "Locked horns" they called it when bowsprits and booms tangled at the wharf. With no bowsprit and a shorter boom, a knockabout schooner like the *Penny* eased the situation somewhat.

The fisherman at the wheel makes up "gangings," short lines which attach the fish hooks to the main groundline of the trawl. He is surrounded by at least thirty-four tubs of trawl. Fishermen normally made up whatever new trawls were required as their vessels fitted out. Existing trawls were overhauled with new hooks or gangings almost constantly, as losses to rough bottom, bad weather, tides, and dogfish were frequent.

Depending on the fish sought—halibut, cod, or haddock—the trawls varied in strength and dimension. While a halibut trawl might be 2,000 feet long, with 150 gangings each 5 feet long, the market fishermen hunting cod and haddock used much lighter gear. Generally, the groundlines were 1,800 to 2,100 feet long, with 500 to 600 gangings, each 1½ to 2 feet long, spaced at 3 to 5 foot intervals. According to a 1913 study, "There is no fixed rule governing the number of hooks on a trawl. Vessels

Bay State

engaged in the offshore fisheries generally use gear with hooks closer together than those employed in fishing on local banks. Captains and crews of vessels entertain different ideas regarding the manner in which trawls should be rigged, and this in a measure accounts for the different styles of gear. . ."[28] Each length of trawl was called a "tub" for the container that held it, except for halibut trawls, which were coiled on a square of canvas known as a "skate."

Schooners generally carried four to six tubs of trawl for each dory. With twelve dories, the *Penny* has at least forty-eight tubs. In a four tub set, she could put 24,000 hooks on the bottom. Since a tub of trawl was worth about $10 in 1910, a schooner's crew had $500 to $700 invested in fishing lines.[29]

The *Ethel B. Penny* was launched in October 1908 by John Bishop of Gloucester. She was built on the model of the McManus-designed *Athena*, very similar to the *W. M. Goodspeed* model of the

Ethel B. Penny

Baiting trawls

same year. Captain Austin Penny took her market fishing for the owners represented by A. M. Watson of Boston. While she usually hunted haddock and mixed groundfish on the South Channel, Georges, and Middle Bank, she went pollock seining in 1909. She also carried a bow pulpit for swordfishing. In 1914, two 37½-horsepower gas engines were added for auxiliary power. The *Penny* survived until 1940.[30]

The *Mary* came in on 1 May, with 35,000 pounds of haddock, 6,800 of cod, 10,000 of hake, and 1,000 of cusk, to find the market still dull. She was modeled by Arthur Binney and built by A. D. Story at Essex in 1912. The *Mary*'s first trip provided an example of the superstition still to be found among some fishermen. Her crew had poor fishing until a bedraggled carrier pigeon flew aboard six days out. They cared for the bird, and the fishing improved. The fishermen later attributed their fine 146,000-pound haul of groundfish to their care of the adopted pigeon.[31] No doubt it was easy to impose an order and establish direct relationships between such occurrences when the conditions of the sea and the activities of the fish were so unpredictable.

The *Mary* was sold about 1920 to Newfoundland, where she fished as the *Mary II* until about 1926. The next year she was back in the U.S. with a 100-horsepower diesel engine. She then disappeared from the registry until 1930. After a rebuilding, she emerged as the *Arthur D. Story* to compete in the elimination races for the International Fishermen's Cup challenge. In 1931 she was fishing as an auxiliary schooner again, with a 150-horsepower diesel. She foundered on St. Pierre Bank with the loss of all seven men aboard, 3 March 1935.[32]

Baiting Trawls at T Wharf, Winter 1912–13

THE WHITE SPARS and freshly painted cabin top suggest that this is a Portuguese market boat, owned in either Boston or Provincetown. Portuguese vessels had a reputation for being well kept. Some of the fishermen seem to identify their trawl tubs with symbols—crosses or a ball and stripe pattern—or differing color schemes instead of the numbers of their dories.

51

Walter P. Goulart

torches smudged their features and gave the only light, paddling bare-handed in boxes of slippery herring and baiting two thousand hooks strung on separate ganges [sic] linked to the most obstinate and cumbersome of lines."

One fisherman stated, "'It's a poor job baitin', you, even with a south wind blowin', but, says I, it's hard, I'm tellin' you, when the snow's flyin' and your fingers bleedin' with the cold, squidjiggin' them hooks.'"[33]

Baiting Trawls Aboard the *Walter P. Goulart,* 1911, and the *Gertrude DeCosta,* about 25 April 1913

THE *Walter P. Goulart* lies in the south dock, far below the caplog of T Wharf at low tide. The *Gertrude DeCosta* came in on 24 April with 11,000 pounds of haddock and 4,500 of cod, and now her crew prepares for sea again.[34]

Gertrude DeCosta

Bait was a precious commodity, and fishermen were often hard-pressed to buy or catch the nearly five tons needed by a schooner for an offshore trip. Because they are baiting in port, these fishermen are probably bound for one of the ledges in Massachusetts Bay, the South Channel east of Nantucket, Georges Bank, or the Nova Scotia coast in search of haddock. Cod and halibut were particular about biting only on fresh bait, whereas haddock accepted salted bait such as herring or clams. For haddock, the fishermen could bait their trawls a day or two before making their first set.

Baiting on the fishing grounds in winter was much less pleasant than in port. An observer who made a trip on the *Evelyn M. Thompson* in 1912 reported watching the "men standing about the cabin till after darkness had fallen and huge oil

52

Fishermen must have become capable of baiting hooks in their sleep. Fishing single dories (one man to a dory), and making two sets a day, each man on these two vessels must have baited up to eight trawls—perhaps 4,000 hooks—each day of good weather. The older fishermen aboard both vessels are taking advantage of the pleasant conditions in port to save themselves some work the first day out. Not so the young fisherman sitting on the *DeCosta*'s main gaff. He gazes longingly ashore now, but at sea he will not sleep until he has done his share of the work. It may take him all night to prepare his trawls for fishing in the morning, but he will learn.[35]

Like dory handling, fish dressing, and vessel handling, baiting was a respected skill among fishermen. A good man could bait a tub of trawl in about forty-five minutes, but of course it was necessary to know which man in which port was the fastest baiter of all. In 1914 some fishermen with money to bet attempted to arrange a trawl-baiting contest on T Wharf to determine whether Gloucester, Boston, Provincetown, or Portland had the best baiters.[36] No formal contest was held, but doubtless there were many informal contests between fishermen on a vessel and between vessels.

The hard usage given a fishing schooner is evident when this view of the *DeCosta* is compared with those of her at Essex seven months earlier. Some of the details of an auxiliary schooner can also be seen. At the stern, her engine exhaust pipe, wrapped with line, passes through the transom, just to the left of the spring-loaded main sheet snubber (which absorbs the shock of the main boom swinging from side to side). The barrel lashed by the wheel may contain gasoline. The dory sails, lashed to the main rigging, are visible on both vessels.

The *Walter P. Goulart* was built by Oxner & Story at Essex in 1904. By 1911 she was owned by Schwartz and Bernstein, clothing merchants and manufacturers of oil clothing in Gloucester. One of the smaller market schooners, she usually made trips of a few days to the ledges of Massachusetts Bay, the Gulf of Maine, and sometimes to Georges Bank. In March 1906 bad weather kept her on Georges for twenty-three days, causing concern at home for her survival.[37]

The inshore ledges were somewhat more sheltered than the banks, and often provided fatter, more marketable haddock and cod. On the other hand, a schooner caught inshore in a storm did not have the benefit of open water to ride out a blow, as the *Goulart* did on Georges (which was dangerous itself in an easterly gale). The alternative, running to port, was the single most common cause of vessel losses. Among the seventy-seven schooners described by Gordon Thomas in *Fast & Able*, about one-quarter were lost while making a landfall. The *Walter P. Goulart* was lost the same way in May 1912. Driven ashore at Yarmouth, Nova Scotia, she was destroyed, while eleven of her crew survived, one drowned, and three missing men later turned up.[38]

Waltham at the South Boston Fish Pier, 31 March 1914

THE *Waltham*'s crew sort out their cod gill nets, having this morning discharged 12,000 pounds of haddock, 7,500 of cod, and 9,000 of pollock taken in their nets during the weekend. The *Waltham* has a triangular riding sail bent on in place of her large gaff-rigged mainsail, allowing nets to be set over the stern without the hindrance of the boom. Gillnetting, she uses her 75-horsepower gas engine more than her sails. Judging from her sternboard, Captain Merton Hutchins, her skipper and owner, is an Odd Fellow. In the background, the stores along Northern Avenue are still unfinished.[39]

The *Waltham* had two careers. She was originally named *Olive F. Hutchins,* for the wife of Captain Hutchins of Cape Porpoise, Maine. Captain Hutchins had been fishing for fourteen years when he had the schooner built at East Boothbay in 1904. Over the previous four years, gasoline auxiliaries had been proving themselves in the mackerel fleet, so Hutchins had a 50-horsepower gasoline engine installed in his vessel, making her perhaps the first auxiliary market schooner.

Captain Hutchins used her fresh fishing, hailing from York and then Kennebunk, Maine, and landing his fares at Portland or Boston. In 1908 Hutchins went fishing for pollock with a purse seine, a

Waltham

method that developed rapidly after appearing about 1905. By 1914, E. & A. Rich & Company, T Wharf fish dealers, owned part of her, and she was regularly landing her fish at Boston.[40]

At 12:30 in the morning of 27 January 1914, she was bound up Boston Harbor with 15,000 pounds of fresh fish, taken during four days on Jeffrey's Ledge. In the darkness, Hutchins and his helmsman watched in amazement as the municipal steamer *Geo. A. Hibbard*, bound down harbor, apparently misinterpreted the *Hutchins*'s lights and, off Castle Island, plowed into her bow, sending her to the bottom. Her crew, most of whom were asleep below, scrambled out just in time and leapt aboard the *Hibbard*. They found shelter

aboard other schooners at T Wharf, where the *Hibbard* deposited them. The T. A. Scott Wrecking Company salvaged the *Hutchins*, patched the gash on her hull, towed her to T Wharf where her fish were unloaded, and delivered her to Richard Green of Chelsea for repairs.[41]

By mid-March, Captain Hutchins had his vessel back, but, having divorced his wife, changed the name to *Waltham*. After gillnetting, he went back to trawl fishing, in December 1914 stocking $1,200 with $43 shares for a week-long haddock trip.[42]

The *Waltham* was still fishing ten years later, handlining in 1925 and fishing out of New York's Fulton Fish Market in 1929. She finally stranded on Block Island, 9 July 1932.[43]

Flaking Down the Seine, South Boston Fish Pier, about 7 August 1914

THE CREW of this auxiliary schooner-boat brings the seine aboard from the seine boat alongside. While probably not a full 1,800-foot seine, it is a large net. It appears to lie in an unruly mess, but the fishermen are actually "making" it—flaking it down—with some care so it can be easily returned to the boat.

The schooner's deck is cluttered with gear. From aft (left) can be seen: the exhaust manifold, wheel box, foghorn, and compass box mounted on an engine housing (the compass must have acted strangely being mounted so close to the cast-iron manifold and engine). On the cabin top are a fish basket, water containers, broom, and probably a dip net for bailing the fish aboard. The barrels are for mackerel, though one contains salt, which the fishermen have been spreading in the folds of the seine to keep it from rotting. Up forward, this schooner has a cuddy cabin and a booby (sliding) hatch over the forward access to the fish hold. Next to the fish baskets amidships is a painted keg buoy, which marks the first end of the seine as it is set.

In this first week of August, mackerel have been schooling in Massachusetts Bay and many sloop-boats, schooner-boats, and powerboats have gotten on fish. This schooner is probably the *Eliza A. Benner* of Vineyard Haven, which has come in with 20 swordfish, as well as 14,000 tinker, 2,000 medium, and 200 large mackerel, taken in Boston Bay. Built at Waldoboro, Maine, in 1900, the *Benner* fished from Waldoboro through 1902; Stonington, Connecticut, through 1909; Edgartown, Massachusetts, through 1913; and Vineyard Haven until her abandonment in 1935. A gas engine was installed in 1905, and was replaced by a diesel in 1925. Her history is representative of many fishing boats that were adapted to changing conditions through auxiliary power, sale to other ports, and seasonal shifts in fish sought.

George Brown Goode proposed that "the purse-seine was undoubtedly a development and extension of the idea of the drag-seine supplemented by that of the gill-net used at sea in sweeping around schools of fish." A pure seine was reportedly in use at Cape Cod in 1839, but it was not until 1860 that the standard design of purse seine emerged.[44]

Seining was expensive. Large seiners often used

Flaking down the seine

two $1,000 seines and two seine boats. The fishing was unpredictable as well. For instance, if mackerel did not show up in the expected numbers, as in 1910, many vessels did not break even for the season. Nevertheless, because of its efficiency, seining quickly became the primary method of catching mackerel.

Seine Boats at Gloucester, 13 October 1912

As the end of a very poor mackerel season draws near, these seine boats are going into storage until the season begins next March. Here, at the head of Gloucester Harbor, Allen B. Gifford carried on the business of building, repairing, and storing seine boats that Higgins & Gifford had established in the 1850s.

The seine boat can be traced back to 1857, according to George Brown Goode. Captain George Merchant of Gloucester had used a whaleboat along with his standard square-sterned seine boat in 1856, and found that the whaleboat performed better. He had Higgins & Gifford build him a twenty-one-foot double-ended lapstrake boat, which became the hit of the seining fleet. Higgins & Gifford's boat was beamier than a whaleboat, with more bearing aft, to support the weight of the seine, while the whaleboat had more bearing forward to prevent the bow from being submerged by the whale line running out. In 1872, the first batten-seam, smooth-sided seine boat was produced and soon replaced the lapstrake model. The seine boat also increased in length as the size of the seine increased, eventually reaching about thirty-eight feet.

Seine boats spent a good deal of time towing behind speedy schooners, were rowed by eight or nine men at high speed to encircle schooling mackerel, and were strained heavily each time the seine was pursed up and "dried in" (pulled aboard to contain the fish in a small section of the seine); yet the boats lasted six or seven years.[45]

According to their catalog for the 1894 season, Higgins & Gifford had produced 1,350 seine boats since 1872, along with trap boats, pulling boats, and small sailboats. With a full crew of fifty men, they could produce ten boats every six days.[46]

Also in storage is the gill-netter *Rough Rider* of Boston. Built at Manitowoc, Wisconsin, in 1904, she was brought east by Captain Edward Widerman in the 1910 migration of Great Lakes gill netters to Massachusetts. These puffing and chugging steam and gas boats, locally known as "smoke boats," fished with a captain and three men, generally in Ipswich Bay. Each day in fall and winter, they delivered their fresh catches of cod, pollock, and other groundfish to market in Gloucester. Much of the catch was then shipped on to Boston.

The small crews of these boats set their nets by hand and hoisted them back aboard with small hoisting engines. One crew fished while another stayed ashore and mended nets.[47]

The little *Rough Rider* continued her mundane but profitable career until stranding in Ipswich Bay, 21 March 1921.

Laying Out and Mending Nets on T Wharf, Winter 1915–16 and Spring 1914

In the first view, lunch hour strollers parade along quiet T Wharf. A couple of boat fishermen lounge against a dory with a brightly painted sheer strake. One schooner is in at the end of the wharf, probably laying up for a few days at T Wharf rather than at the more exposed fish pier in South Boston. At the inner end of the wharf, two boat fishermen lay out a gill net for mending.

In the second view, two fishermen puff on their cigars and patch a net in the afternoon sun. Italian boat fishermen sit and chat in the background.

The net appears to be a mackerel gill net. These untarred nets were about 160 feet long and 19 feet deep, and hung like a curtain at the surface. The corks, spaced about a foot apart, can be seen along the head rope. A few of the small lead sinkers along the bottom edge are also visible. Near the corks is the ball of twine the fishermen are using to repair the three-inch mesh.

Gill nets were walls of netting that allowed fish to swim part way through, only to become snared behind the gills. Gillnetting was a productive but somewhat controversial method used in several fisheries. Mackerel gill netters followed the seiners, finding that gill-netting conditions were best

Seine boats

Laying out a net

Mending a net

on moonlit nights, whereas seining after dark was best on moonless nights. But the gill nets had to be set parallel to the coast, an inefficient orientation for trapping fish but one that kept them from being destroyed by coastal traffic.[48]

Gillnetting for cod, pollock, and other fish in Massachusetts Bay also proved controversial. Norwegian-style gillnetting had been introduced on Ipswich Bay about 1880 for cod. In 1909, Captain John W. Atwood of Provincetown, who had spent twenty years managing a fish plant on Lake Michigan, came home to Massachusetts and demonstrated the Great Lakes method of gillnetting with his steamer *Quoddy*. The following year, the Booth Fisheries Company of Chicago sent a fleet of gill-net boats east, and their success roused local interest.[49]

By 1913, the expanding gill-net fleet began to venture out to Tillie's Ledge, and even to Middle Bank, to set their nets. Those areas, traditionally the preserve of market boats using longline trawls, were the scenes of conflict and destruction of fishing gear as proponents of the two methods competed there in the winter of 1913–14.[50] In another ten years, small inshore otter trawlers were replacing gillnetters in Massachusetts Bay.

4 Under way

DAILY, Boston Harbor offered (and still offers) an ever changing mix of fishing vessels. Though Boston had only slightly more than a hundred fishing vessels, they, and many from Gloucester and other ports, made a total of 3,000 to 4,500 trips annually into Boston during the years depicted here.

After a few days in port, with supplies aboard and crews ready again for sea, they got under way. Depending on wind, weather, and the finances of the vessel, some worked themselves clear of the wharf and set sail immediately, some used their engines, and others required the escort of the little towboats that darted to and fro in the harbor. In the main channel, fishing vessels passed fleets of moored coasting schooners and barges, shared the channel with the same craft and coastal and transatlantic steamers, and in summer passed through flotillas of pleasure craft. Off Boston Light, they set their courses for the fishing grounds and vanished from landbound eyes for a week or more. Already the fishermen would be preparing their gear and themselves for the long days of toil ahead.

When they again threaded their way through the Narrows (until the channel to Broad Sound was completed in the 'teens) and charged up the harbor, the fishermen relaxed as they prepared for a few days ashore. In winter, their toil did not end until they had made that entry. Beating into a northwest wind for a landfall could be as bitter work as any found on the fishing grounds, and running in with a snowy or sleety nor'easter could be

the most hazardous of all. But only blownout sails or ice-encrusted decks gave landsmen a clue to the hardships of fishing.

As these photographs show, in summer the trip up harbor was a cruise, with time for airing clothes, chatting, and watching the sights, all in anticipation of a good stock and a few days of relaxation ashore.

Ruth and Margaret and *Smuggler* at T Wharf, Winter 1915–16

BOTH VESSELS appear to be getting under way. The crew of the *Ruth and Margaret* begin to raise the mainsail. The jib is triced up in the position used to keep it clear of heavy seas. In the harbor, a dory heads toward T Wharf and a harbor ferry passes. On the East Boston waterfront, one of the big Red Star liners lies at a wharf.

Besides the hazards of bad weather and dangerous shores, ocean steamers posed a potential threat to fishermen. Through the nineteenth century, fishing vessels on the banks found themselves in the path of transatlantic traffic. After the Civil War, large iron and later steel steamers sped back and forth across the fishing grounds. Being little hindered by the weather, they often ran blindly through fog and darkness, threatening to run down small fishing vessels without warning. A schooner's foghorn might be heard aboard other schooners, but not on the rushing steamers, over the throb of the engines. In *Captains Courageous*, Rudyard Kipling accurately portrayed the fear and occasional tragedy among fishermen as a liner crossed the banks.

The threat of destruction was much more of a reality than destruction itself, but the possibility of collision remained. Two schooners were run down in 1907, and two more in 1908. The *Fame* of Boston was struck by the British steamer *Boston* and went down with eighteen of twenty men in the Gulf of Maine, and nine of thirteen went down with the *Maggie and May* when she was run down off La Have, Nova Scotia.

In August 1910 the Boston schooner *Belbina Domingoes* was shattered in a collision with the Red Star liner *Samland* on Brown's Bank. The *Domingoes* was fortunate not to founder; as it was, her bow was stove in for at least six feet and her bowsprit and topmasts were lost, but the entire crew survived.[1] The following year, the *Catherine and Ellen* was sunk by a steamer off Cape Cod.

After the *Titanic* disaster in the spring of 1912, the *Fishing Gazette* reported that transatlantic steamer routes would be moved south, away from the fishing grounds.[2] The routes were changed again in the spring of 1913; still, collisions or close calls continued. The Gloucester schooner *Olympia* was struck by a liner and took six of her crew to the bottom off Sable Island in July 1914.

The general adoption of wireless, and later radio telephone in the 1920s, improved navigation and eased the fisherman's thoughts of destruction every time the fog closed in.

The *Smuggler*'s Crew Comes Aboard, T Wharf, Winter 1915–16

THE *Smuggler*'s crew, nattily attired in their blue serge suits, clamber aboard, probably for a trip to Newfoundland for herring. Some of the crew of the *Ruth and Margaret* have already changed their clothes and hoisted the mainsail. Up forward, they have lowered the jib and now cast off the foresail halyards and line up to haul.

The *Smuggler*, the second fishing schooner of that name, was built by A. D. Story for the D. B. Smith Company of Gloucester in 1902. Her first commander was Captain Jerry Cook, a native of Shelburne County, Nova Scotia. In 1884, at age twelve, Cook had come south to work on the first attempt to dig the Cape Cod canal. Remaining in the area, he worked at a fish weir, then shipped on a fishing schooner. After about ten years as a fisherman, he took command of his first schooner. Retiring from fishing in 1917, after about twenty years as a skipper, Cook served as a naval officer during the First World War, spent three years as skipper in the merchant service, and returned to Gloucester to enter the stevedoring and lightering business.[3]

Captain Cook used the *Smuggler* in the fisheries he was accustomed to—mackerel seining in spring and summer, and freighting herring from Newfoundland in fall and winter. By 1908 *Smuggler* was being taken salt banking (using salt to preserve cod caught as far east as the Grand Bank) rather than seining in summer. In 1909 Captain Patrick Shea took her on two salt bank trips between March and October, the first with trawl lines and the second with handlines.[4] By the summer of 1914, Captain Freeman Crowell was using

Ruth and Margaret and *Smuggler*

the *Smuggler* shacking. Generally, shackers remained at sea for two baitings, or about four weeks. Fish caught with the first supply of bait were salted. After a trip to a Nova Scotia port for more bait, unless a suitable amount of bait could be caught at sea, the fish caught on the second baiting were preserved in ice to arrive at market fresh.

The most demanding fishery the *Smuggler* engaged in was the arduous winter herring fishery. The bays on the west coast of Newfoundland produced great quantities of herring, which were caught by local residents dependent on the income they received from supplying the American schooners. Between October and December the herring were packed in salt; between January and March they were generally brought back frozen.

Despite the extremely harsh weather conditions and the frequent damage or loss caused to vessels and crews, the large profits in the fishery kept the vessels returning, often three or more times each winter. But the toll was high. As an example of the conditions, the *Smuggler* lost a fisherman overboard on the return trip in January 1910. She carried such a load of frozen, salt, and pickled herring that thirty tons were stowed in the after cabin. Herring trips did not require a full crew so everyone lived in the forecastle, near the galley. In December 1913, the *Smuggler* arrived from the Bay of Islands, Newfoundland, with a mixed fare of 124,000 pounds of salt cod, 1,065 barrels of salt herring, and 6,700 pounds of smoked salmon.[5]

The *Smuggler* fished out of Gloucester until being sold to Newfoundland owners in 1920.

Crew of the *Ruth and Margaret*, T Wharf, Winter 1915–16

THE *Ruth and Margaret* backs out of the north dock. Having set the main and foresail, some of the crew prepare to raise the forestaysail while others take a break. Several of them have a final chat with their fellows aboard the *Smuggler* while another smokes in solitude. At least one fisherman has not yet changed out of his shore-going suit.

A reporter who spent two weeks with the crew of the *Evelyn M. Thompson* asked:

Smuggler's crew comes aboard

Why do men engage in such a vocation? ... the answer is, primarily, dear reader, because you yourself wish your haddock and your cod on your dinner table when you return home in these days of precious meat. And, psychologically, because from the beginning of the world men have welcomed the opportunity to gamble co-operatively against fate for stakes and for life itself. Just this do these men do. Danger and desire for freedom—those are the attributes that call men to the sea. And those are the attributes, too, that make for individuality. In all his wanderings never has the Spectator discovered sixteen men more individual than these men. From the skipper, thin-lipped, a Socialist, to grizzled old Sam, like another Salters in Kipling's "Captains Courageous," and mountainous big Tom, and bantam weight prize fighter Jack, every one of them differs in mannerisms, methods, and mind from his neighbor. They are accustomed to meet "thick of fogs," days of snow, days of storm, each in his particular manner, simply, without fear. 'I just hauls in me

Crew of the *Ruth and Margaret*

line and keeps an eye on the spot where I see the vessel last, explained Jim. 'I thinks on some things,' drawled big Tom, 'when I's facing death. I ain't got no kids to think on, so sometimes I think on God, and then I's got somethin' comin', ain't I?'[6]

While not necessarily evident to casual observers, profound changes took place in the supply of manpower for New England fishing vessels, beginning about 1850. The use of the trawl line and purse seine, which required strength and skill, discriminated against boys, who could no longer serve their apprenticeships in vessels fishing with handlines. A simultaneous improvement in public education, an unwillingness among many fishermen to have their sons enter such a hazardous occupation, and new opportunities ashore as New England industry blossomed, all reduced the number of young New Englanders entering the fisheries, just as the number of young New England farmers declined. Then, the Washington Treaty, in effect between 1873 and 1885, allowed reciprocal free entry of fish between American and Canadian ports, seriously hampering the New England offshore fleets in the face of cheap Canadian fish. By 1885, only 58 percent of Massachusetts fishermen had been born in the United States.[7]

Soon, the fisheries had become an avenue of opportunity for immigrants. Irish, Scandinavian, and other Northern European men were active in New England fisheries by the 1840s. Portuguese Azoreans and Cape Verdeans, whose fathers had immigrated as whalemen and laborers, especially after a blight in the vineyards, became common in the Gloucester fleet after 1880, and dominated the Provincetown fleet by 1890. But the greatest foreign contingent was the group of Nova Scotians and Newfoundlanders who had begun to appear in New England ports as early as 1830. These "downhomers" made up 11 percent of Massachusetts fishermen in 1885, and increased rapidly thereafter as the markets for Canadian salt fish in the United States and the Caribbean dried up. When *The Fisherman*, a periodical for fishermen, investigated the nationalities of Massachusetts fishermen, as revealed by the 800 casualties in the 1890s, fully 53 percent were British provincials or Canadians. Another 20 percent were of Scandinavian birth, and 2 percent were Portuguese. Only

Ruth and Margaret

The *Ruth and Margaret* in the Stream, Winter 1915–16

THE CAPTAIN of the *Ruth and Margaret*, saving towboat charges, lets wind and tide move her into the stream. The foresail has just been raised, and both it and the mainsail shiver in the breeze. Up forward, the fishermen coil down the fore halyards and prepare to hoist jumbo and jib. One man takes the stops off the jib.

The *Ruth and Margaret* was built by A. D. Story on speculation in 1914. Purchased and fitted out the following year, Captain Valentine O'Neal took her on her maiden trip to Cape North, fresh fishing, in June 1915. She continued as a dory trawler, with a diesel engine added in 1926, until being refitted as an otter trawler in 1928. The *Ruth and Margaret* foundered in Buzzards Bay, 15 August 1948, after thirty-three years of service.[10]

Schooner-boat *Lillian* Getting Under Way, T Wharf, 15 January 1909

WITH THE advantage of wind and tide, the *Lillian* swings away from the wharf as her crew hoists her sails. They are departing two days after delivering 11,000 pounds of cod to market.[11] The foresail is up; the fisherman to starboard sways in on the throat halyard jig to bring it "two blocks" taut. The jumbo and jib gaskets have been cast off, and the other men forward prepare to hoist.

The *Lillian* was a Gloucester schooner-boat, built there in 1902. She received an auxiliary engine in 1909, and she continued to land her fish at Gloucester and Boston until being taken to Gulfport, Mississippi, to fish for red snapper in 1914. She foundered 29 September 1915, taking her crew of seven to the bottom.

These small vessels fished closer to shore than the larger schooners, perhaps going as far as the South Channel or Gulf of Maine. Swordfishing and "rip" fishing—drift handlining from the vessel on the tide rips east of Nantucket—were other activities of the smaller vessels. The fish they landed from the inshore grounds were often fatter, fresher, and more marketable than those brought from the offshore grounds by the large schooners. In

15 percent were American-born. Some masters found the variety of nationalities aboard created a constructive competition among the crew.[8]

The trend continued until, by 1910, the New England fisheries were dependent on the provincial labor supply. When the prospect of free entry of Canadian fish into American markets was raised in 1913, "it was objected that free fish would mean an exodus of the 'downhomers'—the Newfoundland and Nova Scotia fishermen who so largely man our New England fishing fleet and often end by becoming American citizens of an especially valuable type."[9]

Many "downhomers" were in France, fighting the First World War, when this photograph was taken, but the *Ruth and Margaret*'s crew surely reflected the international flavor that can still be found on Boston fishing vessels.

64

1908, the schooner-boat *Cherokee* set a record for shore boats. Fishing for three days off Thacher Island, her crew of three each received a generous $120.[12]

Lillian and *Josie and Phebe* Heading back to Sea, 15 January 1909

As THE *Lillian* swings to port and heads down harbor, the men forward, having raised the head sails, coil the halyards and downhauls. Aft, the rest of the crew cast off the mainsail gaskets and begin hauling on the throat and peak halyards. Perhaps the man standing on the cabin top, nearest the wheel, is the skipper.

Fishing eight dories, the *Lillian* carries ten men. It appears that two or three men on each halyard are enough to set the mainsail. This is not the case on the big haddocker *Josie and Phebe*, getting under way at the same time. She carries twenty-seven men, and probably all of them except Captain Norris and the cook are lined up to haul. The *Josie and Phebe* does not have power, so one of the little Ross tugs has brought her out of the crowded north dock.

The *Lillian* and *Josie and Phebe* contrast in several ways. The schooner-boat makes short trips to the productive grounds near shore where her

Lillian

Lillian and *Josie and Phebe*

Josie and Phebe

fishermen tend their gear single-handedly. The big schooner heads offshore to Georges or the Cape Shore, fishing double dories (two men to a dory). Aboard the *Lillian*, most of the fishermen are relatives or neighbors, some of them may own shares in the vessel, and all of them receive equal shares of half the profits of each trip of fish. It was on these small vessels that the traditional community involvement in fishing vessels survived.[13] The fishermen aboard the *Josie and Phebe* are likely a mixture of Newfoundlanders, Nova Scotians, and perhaps a few Maine or Massachusetts men. Some will remain with the vessel for a number of seasons; others are much more transient. She is owned by a group of investors headed by the attorney Sylvester Whalen, probably made up of fish dealers and other businessmen, who receive one-fifth or one-quarter of the net stock of each trip. The *Josie and Phebe* then represents the urban corporate interests that could afford to fit out and operate one or more large vessels.

Josie and Phebe Under Tow, and "Up Jibs," 15 January 1909

IN THESE TWO VIEWS, the crew finishes setting sail as a tug takes the *Josie and Phebe* down the harbor. Since she is only four months old, the mottled color of her sails suggests they have been "barked" for preservation. Soaking the sails in a mixture of boiled tanbark, and other ingredients depending on personal preference, made them less susceptible to rot and lengthened their working life somewhat.

The *Josie and Phebe* is a large example of the semi-knockabout rig, with a fairly short bowsprit and the forestay brought down inboard of the stemhead. Note the diamond-shaped sheet metal plates to protect the bow planking from being chafed by the anchor stocks.

With a crew of twenty-seven, the *Josie and Phebe* was one of the largest schooners fishing out of Boston. Built by A. D. Story for Sylvester Whalen in the summer of 1908, she was launched in September and finished her maiden trip under

Captain Lawrence Norris at the end of October. Fishing twelve double (two-man) dories, she landed 25,000 pounds of haddock, 12,000 of cod, and 45,000 pounds of hake taken in six days. Four days before this photograph was taken, she delivered 20,000 pounds of haddock, 11,000 of cod, 5,000 of cusk, 5,000 of hake, and 2,000 pounds of pollock to the T Wharf market.[14]

In 1910, Captain Norris in the *Josie and Phebe* had the fourth-best record of the market schooners, with a gross stock of $42,400, and in 1912 led with a $48,000 stock. She continued in the Boston fresh fishery until owners from St. John's, Newfoundland, purchased her in 1918. She was still enrolled at St. John's in 1940.[15]

Henry Fisher entered the second view in a photography contest sponsored by the New York branch of John Wanamaker's Department Store in 1913. His rapid rectilinear lens gave the soft-focus appearance to the composition, which he entitled "Up Jibs."

Schooner *Aspinet* Bound Out, Winter 1911–12

IN A CLOUD of smoke and steam on this gray, wintry day, the *Aspinet*, under the control of a tug, turns from T Wharf to head down the harbor. The tug *Ida M. Chase* scurries across the *Aspinet*'s bow, exchanging whistles with the other tug.

The *Aspinet* was launched 14 March 1908 by James & Tarr. She and her sister, the *John J. Fallon,* were based on Edwin Oxner and Lewis H. Story's *Shepherd King* design of 1904. The *Shep-*

"Up Jibs"

herd King was the second knockabout fishing schooner; with her short bow overhang, she was a distinct departure from the pioneering knockabout *Helen B. Thomas.* Heightened masts compensated for the shortened bow and main boom. The *Aspinet* and *John J. Fallon* were easily identified by their short, blunt "rater" bows, borrowed from "rater" racing yachts. The *Aspinet* made her maiden voyage in April 1908, quickly proved herself first rate, and became known as an excellent vessel in heavy weather.[16]

Captain Jacob O. Brigham, the *Aspinet's* owner and skipper, documented her in his home town, Orr's Island, Maine, but she landed her fares at Boston. Brigham used her market fishing, shacking, and halibuting. He also brought home an occasional swordfish. But for a collision with a barge between Boston and Gloucester in 1909, the *Aspinet's* career proved successful and productive. In 1909 and 1914 her gross stocks were among the top ten earned by the market schooners.[17]

In 1923 the *Aspinet* was sold to owners in Christiansund, Norway. Equipped with a diesel engine, she crossed the Atlantic to serve as a trawler in Norwegian waters. She disappeared from Lloyd's *Register* after 1929.

A Schooner Tows Down the Harbor, early 1914

THE ROSS COMPANY towboat *Sadie Ross* seems to be having difficulty keeping the schooner in line. With a good breeze blowing, the fishermen will quickly make sail and cast off the towline as they overtake the tug. Note the fisherman forehanding the main peak halyard, putting his weight into it to raise the peak of the sail against the breeze. Note that the reef band on the mainsail has been replaced, probably because she has spent so much time on the fishing grounds jogging under reefed mainsail. To the left, the Boston, Revere Beach & Lynn Railroad walking beam ferry strolls across the harbor.

The Ross towboats did most of the towing of

Aspinet

fishing schooners in and out of T Wharf.[18] By 1912, there were seven boats in the Ross fleet, ranging from twenty-seven to forty-nine tons gross. These little tugs were among the smallest but busiest in the harbor. They were constantly towing schooners clear of the wharf, or down harbor if the wind was not fair, and turning around to nudge new arrivals into berths at the crowded wharf.

The *Sadie Ross,* built in 1904, was the largest of the four boats named for members of the Ross family. Captain Joseph Ross had spent his early years in New York towboats. Coming to Boston in 1885, he began building his own fleet of tugs in 1890. The Ross Towboat Company moved to the end of T Wharf in 1900.[19]

Eventually, auxiliary engines and the more spacious facilities at South Boston reduced the fishing industry's demands on the busy little Ross towboats.

Knockabout Schooner Setting Sail off South Boston, 1916

MOST of the crew haul away at the main halyards while the sail billows and thunders. The men to port, on the throat halyard, are about done; those to starboard must now peak up the gaff to the proper angle. The foresail and jumbo are already drawing. The schooner looks very sleek as she knifes through someone's wake, though her properly stowed bait and ice in the after fish pens have her down by the stern a bit. At the stern is a puff of exhaust; on the bow is a swordfishing pulpit.

She is, possibly, the *Evelyn M. Thompson,* built at Bishop's yard in Gloucester for Thomas A. Cromwell of Boston. She was modeled on McManus's *Athena* and built alongside another sister, the *Ethel B. Penny.* Launched in mid-September 1908, she had auxiliary power from the beginning. Captain Charles H. Thompson generally used her as a market fisherman.

On 13 July 1918, while hunting mackerel off Nantucket in a fog, the *Evelyn M. Thompson* ran ashore near Sankaty Light and became a total loss. She was then valued at $12,000.[20]

Schooner tows down harbor

Schooner setting sail

Bunker Hill Monument to Commercial Wharf

Commercial Wharf to India Wharf

70   Fort Point Channel to South Boston

A Panoramic Sequence on Boston Harbor, 15 March 1908

THESE three photographs follow a schooner along the Boston waterfront.

The first view covers the industrialized waterfront from the Bunker Hill Monument to Commercial Wharf; the second view stretches from Commercial Wharf to India Wharf, and the third from the Fort Point Channel to South Boston. This gusty morning, a Boston Tow Boat Company tug steams away from its berth at the end of Lewis Wharf, probably bound down harbor to pick up a tow.

A schooner has also gotten under way. She looks like a fishing schooner of the 1880s, even though she has lazy jacks on her mainsail, uncharacteristic of fishing schooners, except those carrying herring from Newfoundland in winter. Mackerel seiners would fit out in March so perhaps this schooner is headed for Gloucester to take on her seining gear. On such a breezy day, with only a few men aboard, her mainsail is reefed and her main gaff topsail is clewed up.

In the second view, the flock of fishing schooners at T Wharf is visible behind the schooner's mainsail. At Long Wharf lie a white United Fruit Co. banana steamer and a Dominion Atlantic Railroad steamer which makes the run between Boston and Yarmouth, Nova Scotia. Above the tug's steam is the conical dome of the Chamber of Commerce Building (now the Grain and Flour Exchange Building), and just off her bow is the façade of India Wharf.[21]

In the third view, the schooner passes a well-loaded fishing schooner bound in to T Wharf, one of twenty-six schooners that will bring in a million pounds of cod and haddock on this day. With her short foremast and overhanging spoon bow, she is probably the *Muriel* of Boston, with 70,000 pounds of haddock and 20,000 pounds of cod aboard. Built by A. D. Story for the Atlantic Maritime Company in 1904, the *Muriel* and her sister *Selma* were designed by Bowdoin B. Crowninshield. As an Atlantic Maritime Company vessel, she was used haddocking in winter and shacking (salt and fresh fish) in summer. Her winter trips were landed at Boston and the summer ones at Gloucester.

Muriel continued fishing out of Boston until she was torpedoed, 3 August 1918, forty-five miles west of Seal Island, Nova Scotia.[22]

A variety of marine traffic awaits beyond, along the South Boston flats. Visible are a whaleback tanker, two barges, one of which is obviously a cut-down bark or ship, and the training ship *Ranger,* framing the scene at left.

Leonora Silveira in Massachusetts Bay, 17 August 1912

IT IS SATURDAY, 17 August 1912, and the photographer is aboard the Boston-to-Provincetown steamer. Outward bound from Gloucester, the *Leonora Silveira* passes the steamer, displaying the power and grace for which New England fishing schooners of the early twentieth century have become famous. The *Silveira* came into Boston on the thirteenth with 50,000 pounds of haddock, 31,000 of cod, 800 pounds of halibut, and three swordfish. Finding the Boston market glutted with fish, her skipper took her on to Gloucester to deliver her fish to the splitters to be salted or processed.[23]

Many details of a schooner under sail are visible here. The topmasts are intentionally sprung forward to keep the forestays tight. The ballooner (jib topsail) sheet to weather loops slackly across the jib and jumbo (forestaysail) and passes through a bullseye on a becket seized to the shrouds. The ballooner downhaul is slack, as is the sail itself. The fisherman staysail between the masts exerts great pressure on the maintopmast, especially when the main gaff topsail is not set, so a preventer backstay has been set up to take the strain off the maintopmast. The main gaff topsail sheet is visible, running from the furled sail to a sheave at the outer end of the gaff, from which it passes to the deck on the starboard side. The line descending from the outer end of the gaff to the boom is the main gaff downhaul, which is also used as the flag halyard.

The *Silveira,* less than four months old, is departing on her fifth trip to the fishing grounds, but already she shows signs of wear, especially where her dories have scraped her topsides and worn her white boot-top line. One of her dories has a leg-o-mutton sail set, and beneath it two fishermen lean against the dories, watching the Provincetown boat go by. More fishermen are gathered around the trunk cabin, working on their trawls. Note the swordfishing pulpit on her bowsprit.

The semi-knockabout *Leonora Silveira* was launched by A. D. Story at Essex, 5 April 1912, and departed on her first trip on 7 May 1912. She was named for the daughter of her captain and part owner, John Silveira, a native of Pico, the Azores. Her agent was L. J. Costa, Jr., a grocer who represented the largest Portuguese schooner-owning syndicate in Boston, managing seven vessels in 1912. Captain Silveira used his vessel haddocking and shacking out of Boston.

The *Silveira* had a long and fortunate thirty-nine-year career under three names. In 1921 she stranded on Peaked Hill Bars, near Provincetown. Salvaged, she sailed until 1934 as the *Pilgrim* of Gloucester. Stranded in 1934 on Cape Breton, she was abandoned by her owners but salvaged by Canadians and put into service as the *Shirley C.* of Twillingate, Newfoundland. In 1951, soon after being used in the film *The World in his Arms,* she was lost near St. Pierre, loaded with coal.[24]

Clintonia and the Mackerel Fleet at Provincetown, Saturday, 17 August 1912

RAISED MAINSAILS and riding sails or storm trysails keep the vessels' heads into the wind. The nearest vessel is probably the *Clintonia* of Gloucester. Her dory and seine boat, with purse seine neatly flaked for heaving, tow astern. The seine is well salted to keep it from rotting, and the brine from it has stained *Clintonia*'s topsides below the seine roller on the rail. Some of the crew watch the Boston-Provincetown steamer arrive. With a good day in port, someone has hung his clothes to air on the main boom guy. The main topsail clew arrangement, running through thimbles on the leech of the sail, shows up well here. In the distance can be seen one of the weirs, or traps, which supplied both bait and food fish.

Clintonia was one of the noted schooners of the Gloucester fleet. A McManus design launched in

Leonora Silveira

1907, she made her reputation in the mackerel fishery. Under Captain Ralph Webber, she landed the first mackerel of this 1912 season: 250 fish at Lewes, Delaware, in mid-April.[25] Winters, she was used either on Newfoundland herring trips or in the haddock fishery. In both the mackerel fishery and the Newfoundland herring fishery she was known for her long main boom and her fine turn of speed. In 1916 *Clintonia* was sold to Newfoundland. She was lost while carrying salt cod to Oporto, Portugal, in November 1921.[26]

Provincetown Harbor, 17 August 1912

PART of the mackerel fleet swings to a southwest breeze. Because the harbor is shallow, long piers reach out to serve the vessels. The Provincetown-Boston steamer lies at Steamboat Wharf, near several local fishing sloops.

Provincetown harbor was generally a well-protected anchorage. Vessels often sought its shelter in a northeast gale. However, it was open to the south, and in November 1914 three fishing schooners dragged their anchors and went ashore in a southeast gale.[27]

Provincetown maintained an important fishing fleet from the early nineteenth century. By the 1880s, the Portuguese, most of them descendants of Azorean whalemen and other immigrants who had been settling in Provincetown since the 1840s, began to dominate the fishery. In 1877, Provincetown's forty-eight vessels were commanded by six "Yankees," thirty-three Nova Scotians, and nine Portuguese. Six years later, the Portuguese had become the majority, and remained so into the twentieth century.[28]

The Provincetown Portuguese contributed largely to the Boston market fleet. Captains like Manuel Costa, Marion Perry, and Manuel Santos were consistent producers. However, with the

Provincetown Harbor

beginning of the twentieth century, the number of local vessels began to decline, and the use of boats, especially power dories, increased. In 1900, her 650 fishermen manned sixty-two vessels. Six of them went salt banking to the Grand Bank while the rest fished on Georges and the other nearby grounds.[29]

In 1906, Provincetown fitted out only eighteen vessels, including the notable *Rose Dorothea, Jessie Costa, Annie C. Perry,* and the perennial highliner *Mary C. Santos.* Another five lay idle that year, though Provincetown fishermen manned five Boston schooners. In 1914, the town fitted out fifteen vessels over thirty-five tons, fourteen of them schooners, and forty-nine smaller sloops and gas boats.[30]

The Provincetown schooners were usually laid up for the winter. In 1913, the *Mary C. Santos* was laid up at the end of November, after stocking $47,000 for the year. Her fishermen went home with $1,035 shares for ten months of work.[31] During the winter they set trawls from the recently introduced power dories, prepared fish weirs or traps, or dragged for flounder with beam trawls from sloops or gas boats. Flounder dragging began about 1895 in Provincetown. The number of trawls in use rose from 27 in 1898 to 126 in 1908. By 1910 most of the beam trawls had been replaced with otter trawls.

Trap fishing proved lucrative, both for food and for bait. The trap fishermen generally owned their boats in pairs and were paid by the local cold storage plants to tend the weirs or traps set in the harbor and down toward Truro.[32]

By 1912, some of the Provincetown boat fishermen felt that Boston's influence was a bit too close. Provincetown had two freezers until, in 1910, Boston investors financed the Cape Cod Storage and Consolidated Storage freezers for $100,000 each. Soon, the fishermen found they were selling to the freezers (the only local market for their fish) at the accustomed rate, but the freezers had doubled their profit from $3 a barrel (12½ percent) to $6 or $7 a barrel. The fishermen feared a combine would soon control the industry in town, setting prices and using the freezers to control supply. It was another manifestation of the tensions between big business and labor that were working their way into the fisheries.[33]

76

Shenandoah Bound in Through the Narrows, Boston Harbor, 20 August 1912

SAILS SHIVERING in a light breeze, the *Shenandoah* motors through the Narrows, past Lovells Island, on her way up the Boston Ship Channel. Her two seine boats tow astern. Although it has been a very poor mackerel season, the *Shenandoah* has landed fish several times. Today she is bringing in 8,000 pounds of fresh mackerel and fifteen barrels of salt tinker mackerel to a market hungry for mackerel of any sort.[34]

Some of the equipment of a mackerel seiner shows well in this view. There is a safety line strung above the fore crosstrees to support the lookout while watching for signs of schooling mackerel on the ocean surface. Barrels of mackerel stand on deck. The barrel-head box, to hold the heads of open barrels, can be seen on the forward end of the trunk cabin, while the seine roller, which eases the transfer of the seine between the vessel and seine boat stands prominently on the port rail.

The *Shenandoah* was built at Essex by Joseph Story in 1889. Built on the lines of the *J. H. Carey,* with a clipper bow instead of the *Carey*'s "plain stem," she was known to be quite weatherly and a good performer in heavy seas. Her first fifteen years were spent in the service of Petingill and Cunningham of Gloucester, who were later absorbed by William King. In January 1904, she was sold to Captain James Gannon of Boston. He installed a gasoline engine in March 1904, the first time an existing schooner in the fishing fleet was modified for auxiliary power. Captain Gannon used her in the mackerel fishery. In the fall of 1909, Solomon Jacobs, "the Mackerel King," took her shore mackereling.[35]

The *Shenandoah*'s career was cut short only a week after she was captured in this photograph. On 27 August 1912, while fog-bound near Great Round Shoal off Nantucket, she was run down and sunk by the six-masted coal schooner *Addie M. Lawrence.* Captain Gannon and his crew of sixteen took to the dory and seine boat and escaped injury.[36]

Shenandoah

Vessel Coming Through the Narrows, Boston Harbor, 17 August 1912

A MARKET VESSEL roars in past Nix's Mate and Gallups Island with a bone in her teeth. Close-hauled on the port tack, with everything but her fore gaff topsail set, she speeds a trip of fish in to T Wharf. She is probably the *Rose Cabral* of Provincetown, bound in on Saturday afternoon with 10,000 pounds of haddock, 23,000 of cod, 1,500 of hake, and 1,000 of pollock to auction when the market opens Monday morning.[37] Besides her dories, the *Cabral* carries a swordfishing pulpit on her bowsprit. A number of fishermen are gathered aft by the wheel, watching the Provincetown steamer, from which this photograph was taken, go by. The *Cabral* has the white mastheads and booms often seen on the Portuguese schooners.

Designed by Captain George M. McClain and built by James & Tarr in 1890, the *Rose Cabral* was typical of the clipper-bowed vessels often collectively termed the *Fredonia* type after the *Fredonia* of 1889, one of the first deep-draft, clipper-bowed fishermen. The *Cabral* was one of Joseph Cabral's fleet of fine schooners hailing from Provincetown. Like most Provincetown vessels, she usually fished from spring through November, being laid up for several months in winter. In 1914, the last year she appeared in the registry, she was listed with a crew of six, suggesting that she had been reduced to carrying freight. Throughout her twenty-three-year fishing career, she was known as a fast, seaworthy schooner that sailed very well to windward.[38]

Coming through the Narrows

Racing up Boston Harbor

Racing up Boston Harbor, 17 August 1912

THE *Rose Cabral* boils up the harbor under full sail. To port is Spectacle Island. To starboard, a loaded five-masted coal schooner apparently comes to anchor in President Roads. Ahead, it appears that another fishing schooner is approaching Castle Island.

Many skippers found amusement and satisfaction in testing their vessels' abilities and their own skills by racing. There were a number of organized races: "the race it blew" for Gloucester's 250th anniversary in 1892; the Lipton Cup race, won by Provincetown's *Rose Dorothea* during Boston's Old Home Week in 1907; and, of course, the International Fishermen's Cup races between New England and Nova Scotia vessels in the 1920s and '30s. When a new vessel was fitted out for her first trip, it was common for another vessel to give her a run and try her speed, especially if it appeared that she was a fast model.[39] Finally, impromptu races often developed as schooners met on the way to the fishing grounds or as they rushed their catches back to market.

The proliferation of the knockabout model in 1908 brought forth a round of racing fever. Captain Herbert Nickerson felt so confident about his new knockabout *Victor & Ethan* that he challenged the *Rose Dorothea* for the Lipton Cup in 1908. The race never came off, but when the *Victor & Ethan* departed T Wharf six miles behind the larger *Catherine & Ellen* and overtook her in twenty-five miles, the feat was touted in the newspapers. Soon after, going to windward from Provincetown past Peaked Hill Bars, the *Victor & Ethan* outran the knockabouts *Helen B. Thomas*, *Aspinet*, *Matiana*, and, by a slight margin, her sister *W. M. Goodspeed*. In December, Captain Austin Penny wagered a thousand dollars on his new knockabout, *Ethel B. Penny*, against any other knockabouts, but particularly the speedy *Victor & Ethan* or *W. M. Goodspeed*, which were built on a very similar model.[40]

The *Penny*, *Victor & Ethan*, and *Goodspeed* were beaten in turn by the swift little *Frances P. Mesquita* off Highland Light in March 1909. Captain Joseph Mesquita was proud of the *Frances P. Mesquita* and tried her against many vessels. In 1908 she ran from Georges Bank to T Wharf against Boston's noted flyer, the big *Regina*, and lost by only half an hour, even though she was thirty feet shorter than *Regina*. The *Regina* sailed so hard she lost her bowsprit, but such losses were considered merely part of the sport.[41]

Menhaden Purse Seiner *Alaska* off Deer Island Light, Boston Harbor, ca. 1911

THE *Alaska* of Greenport, Long Island, is a long way from home. Perhaps she is stalking menhaden

79

Alaska (detail of panorama negative)

in Boston Harbor, or perhaps she needs supplies. In June 1914 she brought tinker mackerel in to the Boston fish pier, apparently caught as she cruised for menhaden.[42]

The menhaden is an oily fish, seldom eaten. During the eighteenth century, salted menhaden was exported to the West Indies in some quantity as cheap food for slaves on the sugar plantations. In New England it was used for bait and fertilizer. By the mid-nineteenth century, methods of pressing oil from the flesh were developed and factories were built up and down the coast for processing menhaden oil and producing fertilizer from fish meal. By 1875 the products of this industry were more valuable than the products of the American whale fishery. Virginia and North Carolina were at the center of the fishery, though it was followed from Maine to Florida.[43]

In the 1870s steam-powered seiners were introduced in the fishery. Since menhaden were generaly found close to shore, the short trips and bulk carriage made the steamers economically feasible. The fish were still caught by hand-rowed seine boats (visible in the davits), each carrying half the seine. Menhaden fishermen, paid a wage rather than a share, were probably the best-paid fishermen in the 1880s.[44]

Overhead costs were high in the fishery. Coal was a constant expense, seine boats cost $275 to $400, and the seines, replaced each year, cost about $1,000 each. But profits were often high too. The year 1912 was the most productive up to 1914, with 118 steamers and 29 gas-powered boats catching 637 million pounds of fish, worth $2,210,000 off the boat and $3,652,000 in oil and fish scrap.[45]

Built by William Adams & Son at East Boothbay, Maine, in 1881 for the Montauk Oil Company, the *Alaska* was berthed in New Bedford. She had a capacity of 1,300 barrels of fish in the hold, and provided comfortable, imitation oak-grained quarters for two fishing crews totaling twenty-five men. Although the menhaden fishery was relatively profitable, it was also unstable, with companies appearing, merging, and disappearing regularly. The *Alaska*'s record of ownership demonstrates this fact. She was owned in New Bedford until 1887; New York until 1895; Newport until 1897; Greenport until 1913; Newport again until 1919; New London until 1924; Wilmington, North Carolina, in 1925; and Wilmington, Delaware, until disappearing from the registry after 1947.

A Knockabout Bound In, May 1916

A KNOCKABOUT passes Spectacle Island on a broad reach up Boston Harbor. She has her main topmast stepped and her gaff topsail set. The jib seems to be blanketed by the jumbo and is luffing a bit.

With only one dory visible and a pocket spiller boom upright by the port foreshrouds, she seems to be a seiner. It is early for the northern mackerel season, so perhaps she is a pollock seiner. In winter, spring, and early summer a number of vessels went seining for pollock alongshore from Cape Ann east. Pollock seining began about 1905, and increased the landings of pollock significantly. Usually the pollock seiners were gas boats and small steamers, but some schooners pursued the fishery.[46]

It was a tribute to the navigational ability and seamanship of the fishermen that 3,500 to 4,500 vessels laden with fish entered Boston safely each year between 1907 and 1914, regardless of time of day or weather conditions. The ledges outside, the long channel through the Narrows and up between the harbor islands and the flats, the barges and vessels at anchor, and the constant traffic in all directions made Boston a hazardous port to enter. But U.S. Life-Saving Service reports indicate that the Point Allerton Station, which served the outer harbor, was called to assist an average of only six stranded fishing vessels a year between 1 July 1907 and 1 July 1914.[47] While schooners occasionally misstayed and ran aground on the flats or the islands, or tangled with a barge or another schooner, there were surprisingly few serious accidents like that of the *Matiana*. She was running in to Boston during an easterly gale in February 1910, with no visibility in squalls of snow, sleet, and hail. Even in those conditions she might have threaded her way up the harbor, but with a defective compass she wound up on the ledges at North Scituate, fortunately without loss of life. Within the harbor, only the *John J. Fallon*, driving home to market in a New Year's Eve gale, 1914, fetched up hard enough on the flats to grind out her bottom and require rebuilding.[48]

Serious collisions were also rare. The T. A. Scott Company, the major wrecking firm in Boston, salvaged only three schooners sunk in collisions in

the harbor during this period. The *Priscilla* was sunk in 1913, and both the *Olive F. Hutchins* and *Annie C. Perry* went down in 1914.[49]

Gov. Foss off Deer Island, Boston Harbor, Sunday, 23 July 1911

THE *Gov. Foss* drifts lazily up the harbor as a small auxiliary schooner motors out without a stitch of canvas. Nearby, a Ross tugboat, probably the *Betsy Ross*, tows another schooner to sea. The sails of the *Gov. Foss* hang limply; in fact they do not appear to have stretched to shape yet. Even one dory sail is set on this breezeless day. With the mainsail run out so far, the boom guy has been led in to the rail to hold the boom outboard and prevent a jibe. Although it is summer, The *Gov. Foss* does not carry topmasts. Her masts are protected from the chafe of the gaff jaws by sheet-metal plates.

Knockabout bound in

Gov. Foss

The two-month-old *Gov. Foss* is bringing her second trip of fish into Boston, an excellent fare of 125,000 pounds of cod and haddock and 25,000 of halibut, taken on LaHave Bank.[50] The crew is taking it easy, watching the traffic off the starboard bow or enjoying the sun back aft, where skates of halibut trawl line the cabin top. Someone has hung his clothes or oilskins from the jumbo boom.

James & Tarr built this vessel, named for the popular governor of Massachusetts, for Captain Fred Thompson and Captain Lemuel Spinney of Gloucester from a McManus semi-knockabout design quite similar to those of the celebrated *Elsie* and *Elk*. Although she was not particularly fast, *Gov. Foss* was a successful vessel from the beginning. Captain Thompson, a native of Norway, used her for halibut, haddock, and shack fishing, so she landed many trips at Boston. In her first three years she stocked $130,000, a record to that time.

Captain Thompson sold the *Gov. Foss* to the large Gorton-Pew Fisheries Co. of Gloucester in 1916, and two years later they sold her to Newfoundland owners. Returning to American registry in 1922, she fished until stranding and breaking up at Cape May, New Jersey, in 1929.[51]

Schooner off Spectacle Island, Boston Harbor, 4:15 P.M., 12 July 1908

BOUND IN from the fishing grounds, the schooner gives a free tow up harbor to a pair of boat fishermen in their brightly painted Swampscott dory. They are probably Sicilian fishermen who row and sail their dory up and down the harbor, not yet having a gasoline-powered boat. In another two years, no more than one or two of these dories will be left in the Boston fleet.

Schooner off Spectacle Island

Most of the schooner's crew are up forward, watching the land grow near. They may be standing on T Wharf shortly, and will not have to get out their trip of fish until the market opens in the morning.

They are making good time under all sail except gaff topsails. The big ballooner, or jib topsail, sags a bit; the fore topmast stay appears slack. With the fisherman staysail set, preventer backstays have been rigged up for the main topmast.

During this Sunday, twenty-one vessels, including three mackerel seiners, eight swordfishermen, and ten market schooners, arrived to be ready for the opening of sales on Monday morning. The ten market vessels landed 344,400 pounds of ground-fish.[52] This schooner is probably the *Dorothy II* of Salem, arriving late in the afternoon with 2,000 pounds of haddock, 28,000 of cod, and 400 of pollock.

The *Dorothy II* was launched by A. D. Story in 1904, the second schooner of that name built on the model of J. Horace Burnham's *Boyd & Leeds* design of 1894. In all, seventeen schooners were built to this fast, seaworthy design, which was quite similar to the Baltimore clipper hull.[53] *Dorothy* went to Pensacola, Florida, as a red snapper fisherman in 1910. She foundered off the entrance to Mobile Bay, 6 July 1916, with the loss of seven of her crew of nine.

Athena Tacks up the Inner Harbor, Sunday, 23 July 1911

Athena is framed by the heavy smoke of the Boston-to-Yarmouth liner *Prince George,* which has just left Long Wharf and now accelerates down the harbor. Between the *Prince George* and the coasters anchored on the South Boston flats is a dredge, working on the long project to improve the channel in Boston Harbor.

Athena is bringing in 51,000 pounds of haddock, 2,500 of cod, and 3,500 of hake for Monday's market. A Thomas McManus design, *Athena* was the prototype for the *Ethel B. Penny* and the *Evelyn M. Thompson*, and was quite similar to the *W. M. Goodspeed*. She was built by Bishop at Gloucester in the spring of 1908 for the Boston fleet. Under Captain Edward Forbes, she had the fifth-best stock of 1909, totaling $32,000.[54]

In the fall of 1911 a 100-horsepower gasoline engine was installed in preparation for the *Athena's* trip 'round Cape Horn to join the halibut fleet at Seattle. In November, a few days after the *Victor & Ethan* departed for Seattle, Captain Edward Brewer got *Athena* under way, as vessels in the harbor saluted her with flags and whistles. Both schooners made the passage in about four months.[55]

With an increase in the number of Pacific-built halibut schooners, which had engines, pilothouses, and power equipment, the New England model schooners lost their prestige in the Northwest. *Athena* returned to Boston about 1915 and remained in the Boston fleet until she was sold to Newfoundland owners after 1928.

A View Down Boston Harbor, Sunday, 23 July 1911

THE knockabout schooner *Athena* beats up the harbor to T Wharf. Beyond her, largely on filled land, lies the developing industrial area of South Boston. The new fish pier is under construction at the right of the photograph, under the smoking stack. At the far left is Castle Island, long a park providing recreational relief in the summer for the working families who could not leave the city. Several empty coasters lie at anchor in the background. In the foreground, a five-master laden with coal waits to be towed to a berth to discharge her cargo.

These huge coal schooners (and increasingly, strings of barges) provided the fuel to produce electricity, heat homes, and power New England industry.[56] They also posed occasional hazards to the fishermen. While perhaps three times as long as a fishing schooner, they might carry a crew less than half the size. Leaving Philadelphia or the Chesapeake, they headed up the coast and through Vineyard Sound, where they waited for a fair wind. Proceeding through Nantucket Sound, they rounded Monomoy Point, skirted the deadly curve of Cape Cod, and joined the traffic bound across

Athena

View down Boston Harbor

Massachusetts Bay to Boston. From Nantucket on, they passed through waters frequented by the market fishermen and seasonal mackerelers. And like the transatlantic steamers, they were unwieldy vessels trying to make time, even in congested waters.

Collisions were not common, but they did occur. The worst area for collisions lay between Nantucket and Monomoy Point. Fog was often a hazard here. In August 1912, the six-master *Addie M. Lawrence* ran down and sank the mackerel schooner *Shenandoah* in a fog off Nantucket.

Night also posed a problem. The *Margaret Dillon* lost her bowsprit to a five-master one November night in 1913. A year later, the *Hattie Heckman*, anchored off Handkerchief Shoal near Monomoy, had her bowsprit raked off by a four-master.[57]

Note the long driver (after mast) boom on the five-master. With all fifteen sails (or nineteen if she carried staysails) set she would have been quite swift. In fact, the schooners that clipped the *Dillon* and the *Heckman* were gone before they could be identified.

Crowd in the north basin

5 Landing Fish

APPROACHING T Wharf, fishing captains could judge the market conditions. Many vessels in port usually indicated a glut of fish and low prices, while vacant slips often represented premium prices and a fine stock (gross return), even for a small catch.

Until early 1909, docking schooners were met by fish buyers dickering for their fares from the caplog of the wharf. This confused and inefficient system was terminated by the institution of the still-existing auction system of the New England Fish Exchange.

From schooners and sloops came haddock, cod, halibut, hake, pollock, cusk, mackerel, and swordfish; from the otter trawlers came haddock, cod, and flounder; from boat fishermen came lobster, crabs, herring, and assorted groundfish; and by rail, aboard coastal steamers, and in small transport craft from Maine and Cape Cod came lobster, trout, salmon, sardines, and other processed fish. Between 1907 and 1914, the annual landings of unprocessed fish at Boston hovered near or above 90,000,000 pounds.

A Crowd in the North Basin, Thursday, 16 February 1911

AT LEAST sixteen schooners are visible in the north dock; in the last two days thirty-two schooners and two steam trawlers have arrived, and today seven more schooners and a steamer will come in. Arrivals since Tuesday include the *Annie C. Perry, Mary C. Santos,* and *Jessie Costa,* hustlers from the Provincetown fleet, and the noted offshore schooners *Regina, Rex, Elmer E. Grey, Esperanto, Mary F. Curtis, Elsie, Onato,* and *Slade Gorton.* Despite the number of vessels in, prices have remained solid: two cents and more a pound for haddock, and over four cents a pound for both cod and hake.[1]

On the left, above two idle power dories of the Italian "mosquito fleet," the *Flora S. Nickerson* lies at Commercial Wharf. She came in a week ago from a brutal three-week trip to Georges Bank, during which her mainsail was blown out of the bolt ropes as she lay hove to, and her rudder box and steering gear were smashed by a boarding sea. She has not yet received her new mainsail.[2]

Just aft of the *Nickerson,* the Plant Line steamer *A. W. Perry,* an 1897 product of Belfast, Ireland, has just arrived from Halifax and still has steam up and two pennants flying. In front of the *Perry,* a new arrival is entering the dock with a steaming little tug at her starboard quarter. She is probably the *Mary E. Silveira,* coming in with 8,000 pounds of haddock, 500 of cod, and 800 of hake.

The clipper-bowed schooner alongside T Wharf is almost certainly the big *Regina,* queen of the Boston fleet, and reputedly the fastest schooner out of Boston. She arrived two days ago from offshore with 17,000 pounds of haddock, 4,000 of cod, and 10,000 of hake. A little more than a week ago, the *Regina* was hove down by the force of the gale on Georges Bank, lying with her mastheads nearly in the water. Now, a man is aloft, apparently checking the blocks for the main peak halyard.

The *Regina* has also brought in some of the trawls left set by the crew of the *Josephine DeCosta* when they were forced to run in to Provincetown for shelter during recent gales. Because of two weeks of heavy weather, the *Mary C. Santos,* which also arrived Tuesday, had to seek shelter in Provincetown Harbor as well. But her resourceful captain, Manuel C. Santos, took her out three times during lulls in the storm to make single sets and then return. He shipped the fish to Boston by rail, making $2,200 for the three sets.[3]

Despite the often wintry weather, Lent was the best season for haddockers. The demand for fish in Catholic Massachusetts was met by the landing of four to six million pounds of haddock each March between 1907 and 1914, and February landings were close behind. On two consecutive days in February 1909, a total of 100 vessels put in to T Wharf with two million pounds of fish; but perhaps the busiest day was 17 March 1909, when 61 vessels arrived.[4]

However, statistics from the U.S. Bureau of Fisheries annual reports indicate that, on a seasonal basis, summer (July, August, September) was the busiest time on T Wharf, with an average of almost eleven hundred landings of mackerel, swordfish, halibut, and a variety of groundfish, brought in by both vessels and boats. Fall (October, November, December), with the end of mackerel and the beginning of the haddock season, was almost as busy, with just under eleven hundred landings. The haddock fishery and the beginning of the cod, mackerel, swordfish, and halibut seasons brought just over nine hundred trips of fish to T Wharf in the spring (April, May, June). During winter (January, February, March) the haddock fishery was predominant, but despite a steady demand, a reduced fleet operated, with an average of just under nine hundred landings.[5]

The *Flora S. Nickerson* was one of Thomas McManus's Indian Header designs, built by James & Tarr in 1902. Managed by Thomas A. Cromwell, she fished out of Boston through 1911. Sold to owners in Fortune Bay, Newfoundland, she had disappeared from the British registry by 1924. The *Nickerson* and her sister the *Matchless* had the rounded "rater" bows of the so-called Indian Headers (because the first few of the design had Indian names), but had slack bilges, which allowed them to heel over in blowing weather, rather than standing up and stressing their spars and rigging.[6]

The *Mary E. Silveira* was built at Gloucester in 1904, and served as a market fisherman out of

Boston until being sold to Gloucester owners in 1914. Apparently she soon went red snapper fishing, as she foundered in the Gulf of Mexico with the loss of all nine aboard, 16 August 1915.

The *Regina* was built by A. D. Story in 1901 for William Emerson of Boston. Captain Jeremiah Shea skippered her in the fresh fishery through most of her days out of Boston. She was laid up at Long Wharf from late 1911 to late 1913, and then sold to owners in Bucksport, Maine. They apparently used her as a coaster, as her entries in the *Merchant Vessels List* give crew figures too small for a fishing schooner (nine in 1915 and five in 1916). She was returned to fishing with a crew of twenty-two during the profitable years of 1917, 1918, and perhaps 1919. The *Regina* burned, 22 December 1919, at St. Vincent, Cape Verde Islands. She had thirty people aboard (none were lost), suggesting that she was another fishing schooner that finished her career as a "Brava packet," running between New Bedford and the Cape Verde Islands.[7]

Schooner Coming in to T Wharf, 1:15 P.M., 4 September 1908

BRINGING a schooner in to T Wharf without a tug could be tricky business, especially with a stiff breeze and the basins full of vessels. Here, on a quiet afternoon, a medium-size schooner ghosts in under a lowering mainsail. One fisherman in a straw hat is out on the bowsprit, tying off the neatly furled jib.

Nine schooners will land 207,000 pounds of fish today. The white mastheads and gaff suggest that this is one of the Portuguese schooners, perhaps the *Emelia Enos* of Provincetown, delivering 16,000 pounds of haddock, 10,000 of cod, 2,000 of hake, and 1,000 of pollock.[8]

The *Emelia Enos* was built by James & Tarr at Essex in 1902 for the Provincetown fleet. After twelve years, she was sold to Pensacola to serve as a red snapper fisherman. Twenty-seven years later she was abandoned as unfit, after thirty-nine years at sea.

Where is she coming from? The great majority of Boston fish arrived fresh, from the nearer grounds: the South Channel east of Nantucket, Middle Bank, Georges Bank, the Cape Shore (the Nova Scotia coast near Cape Sable), Brown's Bank, the Chatham Shore, Jeffrey's Ledge, and the ledges along the North Shore and the Gulf of Maine. Schooner-boats, sloop-boats, and power-boats brought in most of the shore fish, and the smaller schooners often ventured as far as the South Channel and Georges Bank.

After about 1912, the increasing number of otter trawlers raised the proportion of fish landed from the South Channel, Nantucket Shoals, and Georges Bank, the areas where otter trawling was best suited. Schooners could fish on both smooth and rocky bottom, and frequented the Cape Shore, Brown's, Middle, and Georges Banks, and the South Channel, holding their own against the otter trawlers through the 1930s.[9]

Discharging Fish on a Busy Day in the North Dock, Winter 1911–12

SCHOONERS appear to be five deep at the wharf this afternoon as the little tug *Irving R. Ross* steams in with a new arrival. The vessel unloading is a single dory haddocker, carrying fourteen dories. The dory sails, which single fishermen used to propel their dories while setting trawl, can be seen lashed to the main rigging. While some of his shipmates discharge fish, the nearest fisherman scrubs down the vessel with seawater and a broom. Notice the wear in the bait cutting board on the cabin top. One of the fishermen on the gurry box has left his pipe on the loosely furled mainsail.

Once a vessel arrived at the wharf and the fish were sold, her fishermen had another eight hours of labor discharging fish. Traditionally, all fish except the large, flat halibut were hoisted out of the hold in baskets. Several men loaded the baskets in the hold, others hauled the tackles to hoist the baskets up, and a couple of men walked them to the wharf to be weighed. Two such teams are working in the photograph, while apparently bantering with men on the wharf.

The *Elizabeth C. Nunan* was the first schooner to carry a gasoline-powered hoister on deck. In August 1910 she used it to unload 80,000 pounds

Coming in to T Wharf

of fish in only four hours. In 1914 the *Saladin* was equipped with an electric deck hoister. Gradually, these deck hoisters, successors to the steam donkey engines of the late nineteenth century that allowed the big coasters to function, proved their worth and became standard equipment on many sailing fishermen later in the 'teens.[10]

Weighing Out Fish, 15 January 1909

FISHERMEN in their oilskins and representatives of a fish dealer gather around the scales as fish are hoisted out of the schooner's hold. Fish, coming out of the hold in baskets, are dumped into the boxes on the scales, under the eyes of the captain or his representative, the purser, and a representative of the fish dealer who has purchased that part of the trip of fish. Fish are then transported in carts to the dealer's processing rooms.

89

Traditionally, dealers either made arrangements to buy a schooner's fare before she sailed, or else offered a price when a vessel arrived, subject to changing conditions during the day. For instance, if the *Belbina Domingoes* came in alone with 30,000 pounds of cod and haddock early Monday morning, she would receive a fine offer; but if thirty-five other schooners had arrived by noon the offer would be reduced, with very little recourse for the captain, except holding out for another day or taking his fish to Gloucester to be split.

Discontent with the situation led to a meeting between the Fishing Masters Association and the Fresh Fish Dealers Association in November 1908. Less than two weeks after this photograph was made, there was a drastic but beneficial change in the method of delivering fish: the New England Fish Exchange was opened to provide an auction form of marketing for fish. Thereafter, rather than being approached by dealers on the wharf, each captain brought an account of his cargo to the auction room at the outer end of the wharf. Transactions began at 7 A.M. and continued at signaled

Discharging fish

Weighing out fish

times during the day. Dealing through the Exchange, a captain could count on having a dealer take the amount of fish agreed upon at the price agreed upon. A dealer could count on receiving the amount and condition of fish that he had offered to purchase. A system of records and receipts was introduced. Disputes could be arbitrated, and the Exchange agreed to "do all in its power to advance the interests of the fishermen in a legitimate way that does not seriously conflict with the interests of the dealers and others." While fishermen at first complained about the one percent per fare charge for dealing through the Exchange, they agreed that the Exchange improved their ability to sell their fish and get to sea without delays. The New England Fish Exchange operates to this day on the South Boston Fish Pier.[11]

Note the lack of maintenance on the schooner, visible in the dangling "Irish pennants" and broken ratlines. Fishermen generally did not consider routine maintenance to be among their duties, so it was only done when financed by the owners.

An Icebound Schooner at T Wharf, 16 February 1911

A CROWD of fishermen and spectators chat and inspect the coating of ice on the schooner as fish are hoisted out of the hold and weighed on the wharf. Astern, the steam trawler *Ripple* loads ice. The schooner is a semi-knockabout, with short bowsprit and the forestay brought down inboard of the stemhead. She is almost certainly the famous *Elsie*, the only semi-knockabout to land fish today. She is discharging 25,000 pounds of haddock, 8,000 of cod, 8,000 of cusk, and 1,000 of halibut.[12]

The cart holds a load of groundfish, probably cod or haddock. These are average-size fish, perhaps three to eight pounds. Occasionally, a monster cod that had passed up the lure of bait for years was brought in on a trawl. The *Athena* landed a 120-pound cod, with a head as large as that of a Newfoundland dog, on Middle Bank in 1910, and the *W. M. Goodspeed* took a six-foot, 90-pounder, which registered 65 pounds dressed, on the same bank in 1914.[13]

Rules on fish quality were set in 1909 as part of the New England Fish Exchange attempts to improve conditions. Fishermen were required to dress the fish on deck promptly, rather than letting them sit and get "sun cooked" through the day, and to thoroughly clean and gill them to prevent rapid deterioration. Salt fishermen were reminded to use plenty of salt, so the fish did not come in heavy with moisture (for a better price) but poorly preserved. Shack fishermen were warned that all of the fish taken on the first baiting had to be salted and only those taken with the second baiting could be iced.[14]

By 1914 it was estimated that a hundred million pounds of groundfish were landed at Boston each year. Haddock was the primary fish, comprising about 45 percent of the landings. Cod was second at 25 percent; followed by hake, 16 percent; pollock, 10 percent; cusk, 2½ percent; and halibut, one-half of one percent.[15]

By 1910, 51 percent of Boston's fish rode the New York, New Haven & Hartford Railroad out of town for distribution throughout the Northeast. The name train "Flying Fisherman" was initiated in 1913 to rush fresh fish to New York overnight. Thereafter, fresh fish from Boston graced tables in Philadelphia, or even Chicago and St. Louis, within two or three days of being hoisted out of a schooner's hold at T Wharf.[16]

The *Elsie* was built by A. D. Story for the Atlantic Maritime Company of Boston in 1910. Her design was based on Thomas McManus's model for the *Oriole*. While owned by Atlantic Maritime, she was generally used haddocking, shacking, and halibuting. When Atlantic Maritime

Icebound schooner

sold her to Captain Alden Geele of Gloucester in 1916, she went salt banking for three years before spending two years under British registry with Lunenburg, Nova Scotia, owners. She is best known for her valiant but unsuccessful defense of the International Fisherman's Cup against the larger *Bluenose*, just after returning to American registry in 1921. With an oil engine installed in 1924, she went fresh fishing and carried herring from Newfoundland. Her engine was removed about 1930. Sold to Newfoundland owners in 1934, she foundered in the Gulf of St. Lawrence in June 1935.[17]

Arethusa in the North Dock, Winter 1912–13

THE *Arethusa* may be about finished discharging a trip of fish. One fisherman stands by the main shrouds, either waiting to receive an empty fish basket or chatting with spectators on the wharf. Behind him can be seen the "checkerboards," low pens designed to hold the large, flat, slippery halibut until they could be dressed. Beside the same fisherman is one of the large tubs used for washing fish before they were stowed away in the hold.

Arethusa, the first knockabout schooner built by James & Tarr, was launched at Essex in September 1907, and remained the largest and fastest knockabout until the *Catherine* was built in 1915. She was known primarily for the many successful trips she made for salt cod under Captain Clayton Morrissey, renowned in his own time as "one of the greatest Grand Bank skippers that ever trod a deck."[18] However, after the salt banking season, Morrissey and other skippers took her out for fresh fish. On 9 April 1913, Captain Joshua Stanley, making a last trip before retiring from the sea, brought her in to T Wharf with 186,000 pounds of groundfish, which proved to be the largest single fare ever landed at T Wharf. On her first trip to T

Arethusa

Isinglass factory

Wharf, in February 1912, she had landed 110,000 pounds (85,000 pounds of cod) caught in ten days, for a return of $3,500. In the winter of 1913, between 20 January and 9 April, the *Arethusa* landed six trips of fish in Boston, almost one every thirteen days. Her total catch was quite respectable: 214,000 pounds of haddock, 185,000 of cod, 1,400 of halibut, and 89,500 of hake, cusk, and pollock.[19]

The *Arethusa* had a notable career, even beyond the consistently large stocks of fish she landed. She once outran a Canadian fisheries patrol steamer, and later narrowly escaped destruction on the shoals of Sable Island. She was a well-known rum runner for a time after 1921, and was finally lost in 1929.[20]

One night in April 1912, the *Arethusa* avoided premature destruction in an imminent collision with a steam trawler by employing her new electric searchlight. John Hogan, son of a Gloucester skipper, modified an automobile headlight for use on board ship, and his design was quickly adopted. The skippers of the *Frances S. Grueby* and *Laverna* experimented with the electric lights in January 1912. By September, most of the schooners hailing from Boston and Gloucester carried them. The electric lights were safer and more convenient

than the standard kerosene flareup lights, particularly for a vessel threatened with a collision or searching for a lost dory.[21]

Isinglass Factory, Rockport, Massachusetts, Summer 1912

SEVERAL MEN lay out wire screens covered with fish sounds (swim bladders), like those in the foreground. Note the booby hatch from some vessel being used as a support for the homemade drying screens in the immediate foreground.

In the name of profit and the spirit of conservation, several parts of edible fish were saved beyond the flesh; for example, cod livers were often saved for their oil, which was used by the tanning industry. Fish sounds or swim bladders, which regulate the specific gravity of fish, allowing them to rise or sink in water, also had commercial uses. The industry was a small adjunct of both the Gloucester and Boston fisheries. Gloucester salt bankers removed the sounds as the fish were cleaned, and they traditionally became a perquisite of the cook. In 1908, Suffolk County (Boston) contributed 87 percent of the 73,000 pounds of sounds recorded, for a total value of $3,100. Large hake caught in the deep waters off Nova Scotia provided the most suitable sounds.[22]

At the isinglass factories, raw sounds were air dried. The dried sounds were chopped and rolled into sheets, then cut into ribbons and dried again. These ribbons of isinglass were used to clarify beer and wine, and as an ingredient in gelatin.[23]

A Schooner-load of Herring at Long Wharf, 15 January 1909

THE NEW little knockabout *Mary J. Beale* has come in with a cargo of herring from Eastport, Maine. There are at least eight barrels fore and aft,

Schooner-load of herring

along with a scale to weigh them as they are hoisted out of the hold. The foresail has been unbent from its gaff and the gaff is hoisted as a cargo boom. In the background, a number of market schooners sit in the south dock. One, with main topmast still standing despite the winter weather, has her sails partially hoisted to dry.

The *Beale* appears to have a brightly striped rail. While her low trunk cuddy dates back to the early-nineteenth-century pinky design, she is quite a modern sardine carrier, probably reflecting the influence of large vessel designs on small, outport vessels. Built in 1908, the *Beale* represents a modification of the sardine carrier with the addition of an auxiliary engine (already becoming common in the fleet), and the knockabout bow, which was gaining popularity in the major ports.[24]

The *Mary J. Beale* has probably come in with one of the last trips of Maine sardines of the 1908 season. Like mackerel, the herring were unpredictable, often being in short supply. Law limited the use of herring weirs to April 15 through December 1. Herring were transported to canneries in Eastport, Lubec, and other Maine ports, where most were canned as sardines. Vessels like the *Beale* distributed the sardines.[25]

Since barrels are being unloaded, rather than boxes of sardine cans, it is possible that the *Beale* has landed salt herring, either to be smoked or to be used as bait. Newfoundland herring were carried to Gloucester and Boston in the winter to be smoked, and small herring from anywhere were much in demand for bait. The large fleet of vessels fishing with handlines and trawl lines was dependent on a steady supply of bait. The development of large freezers, like the Quincy Market Cold Storage plant visible in the background of some of these photographs, reduced the seasonal fluctuations in the availability of bait.

Documentation lists reveal the decline of the *Mary J. Beale.* In 1914 she was sold to fish out of Southwest Harbor, Maine; in 1920 to Boston; in 1923 to Gloucester; in 1925 to Providence; and in 1926 to New Haven. She foundered off Fire Island Light, 11 December 1926.

6 Passing of the Fishing Schooner

THE graceful schooners that form so large a part of this work were becoming obsolete even as they reached the peak of their design. Most of the technologies that superseded them were adopted or perfected during the period encompassed by these photographs. The fleet changed rapidly: the twenty-three schooners pictured here that were sold to Canada or out of the fisheries spent an average of only twelve-and-a-half years fishing from New England ports. While schooners remained active for another twenty years, these photographs foreshadow the end of sail power.

The changes came rapidly. Competition from otter trawlers was nonexistent until 1905, and only one such vessel operated between 1905 and 1910. Thereafter, the number rose to nearly fifty by the end of the First World War, and the method was definitely established.

Gasoline power became almost a necessity in the schooners, beginning in the mackerel and swordfishing vessels between 1900 and 1910. Diesel or crude oil power became a reality, tentatively with the Blanchard engines of the *Bay State* and *Knickerbocker* of 1912, and practically with the *Manhassett*'s Nlseco diesel in 1914.

At the same time, the knockabout schooner design, which appeared in 1902, received its greatest boost in 1908, when the number in existence tripled. By 1912, the design combining the best features of the knockabout schooner and auxiliary power had been created, and the apex of the working New England fishing schooner was reached.

The quality and longevity of New England fishing schooners have long been debated. Some claim the schooners were the strongest, most durable, and all-around best working sailing vessels built. Others claim that they were both quickly and cheaply built, and that their service was correspondingly short. The average life span of the thirty-six sailing vessels pictured here that can be traced was twenty-two years. But it is a telling fact of the rigors of the fisheries that only nine of them survived long enough to be abandoned as unfit for service. Two burned; two sank in collisions; four were destroyed in the First World War; nine stranded on shoals or inhospitable shores; and ten foundered in the open sea, several at an advanced age.

The 1920s saw the end of all-sail powered vessels, the *L. A. Dunton* of 1921 probably being the last strictly working schooner of large size entering service without auxiliary power. Vessel types such as those represented by the *Mary F. Ruth*, the Bay State Fishing Company trawlers, and perhaps even surplus World War One submarine-chasers, contributed significant features to the designs of new vessels, but even into the 1940s, elements of the classic schooners could be found in new diesel trawlers.

Dory Trawler on the Delaware River, 21 September 1907

PHILADELPHIA dealt more in oysters than in finned fish, but the city did have a few fishing schooners. M. P. Howlett was probably the largest

Trawler on the Delaware River

dealer, and this schooner is likely his namesake.[1] She was built by A. D. Story in 1901 as the *Jennie & Agnes* and fished out of Boston for three years before being sold to Howlett. She disappeared from the records after 1910, perhaps being sold to Cuba or Nova Scotia.

She is typical of the moderate-size clipper-bowed schooners built in large numbers around the turn of the century. This model of schooner was largely superseded by the round-bow, "Indian Head" type, which was followed by the knockabout design. Fishermen and owners were quick to try the latest successful design when the economy warranted. The average age of Boston schooners in 1914 was nine years, an age which roughly corresponds to the appearance of the round-bow schooner and the knockabout (almost one quarter of Boston schooners were knockabouts in 1914).

While hard usage and disasters took their toll, more schooners were disposed of by sale. Gloucester and Boston occasionally exchanged vessels, but the usual pattern of sale was from the major ports to the outports. Schooners might be sold to Prov-

incetown or Maine (or Philadelphia), but most went to the red snapper fishery of the Gulf of Mexico and the expanding Nova Scotia and New-foundland fisheries. A resurgence of the Canadian fisheries around 1910 and the depredations to the fishing fleets that occurred during the First World War both contributed to a high-priced demand for surplus schooners in the Maritime Provinces, which was met by a steady flow of eight- to twelve-year-old American schooners.[2]

A Typical Mixture of Market Boats in the North Dock, about 25 April 1913

In 1914 there were about 475 vessels landing their catches in Boston, including 285 schooners, 150 sloops, powerboats, and small craft, and 26 steam vessels. This photograph presents a cross-section of those vessel types, some 45 of which have arrived in the past three days to find low prices prevailing.[3]

In the background is one of the three Bay State

Fishing Company steam trawlers that came in during the last two days, and a knockabout schooner. The knockabout schooner *Pontiac,* schooner-boat *Dixie,* and sloop-boat *Lillian* occupy the middle ground. In the foreground, a Sicilian fisherman overhauls his trawls. He will probably use the fish in front of him for bait. Under his feet is his jug of water; on the other side of the cuddy hatch is the bilge pump. The engine is housed in the cuddy cabin but, for use in case of engine failure, long oars are lashed atop the cabin.

The sloop-boat *Lillian* and schooner-boat *Dixie* also represent the inshore fisheries. While the shore fisheries were declining by the end of the Civil War, economic depression in the 1870s, and the 1871 Washington Treaty which allowed duty-free entry of cheap Canadian fish into American ports, beginning in 1873, led to construction of numerous sloop-boats and schooner-boats rather than more expensive full-size schooners. A few continued to be built into the twentieth century.[4]

Built at Friendship, Maine, in 1910, the *Lillian*

Typical mixture of market boats

is a variant of the classic design of the Maine sloop-boat known now as "Friendship." Like many of the sloop-boats, she has a small gasoline engine. Her crew of four uses her on the ledges around Cape Ann, either with trawl lines or gill nets. On this trip she has delivered 700 pounds of cod. She has either a table for baiting and dressing fish or a grating for flaking down her anchor cable on top of the cuddy cabin. She also has washboards fitted so that she will not lose her dories overboard. The *Lillian* was finally abandoned in 1941.

The schooner-boat *Dixie* came in on 23 April with 2,700 pounds of cod. Some of her crew of six are aboard now and have set the jib and one dory sail. Built at Beverly, Massachusetts, in 1880, the *Dixie* is already thirty-three years old, but she will go on for another twenty years before being abandoned.

The *Pontiac* has delivered a small catch of 150 pounds of haddock, 1,800 of cod, 8,000 of cusk, and 19,000 of hake. She was the first McManus short-bow knockabout, the fourth knockabout built, and the first constructed at Gloucester. Launched by John Bishop in 1906, she had a very productive ten-year career, finally stranding on Handkerchief Shoal in October 1916.[5]

Under Captain Enos Nickerson, the *Pontiac* was used mackerel seining in spring and summer and market fishing the rest of the year. In 1909, her $16,700 mackerel stock was second behind that of the *Mary E. Harty*. Her fishermen received $402 shares for the season.[6]

Later, under Captain Ernest Parsons, the *Pontiac* produced consistently as a market schooner. Captain Parsons, born in Rencontre, Newfoundland, in 1886, was an able seaman when he first shipped out on a fishing schooner at age eighteen. By 1911 he commanded the *Ethel B. Penny*, giving her up that December to replace Captain Nickerson in the *Pontiac*.[7]

Captain Parsons was known as a driver. During the terrible weather of January 1914, he brought the *Pontiac* in with 60,000 pounds of haddock, 7,500 of cod, 4,000 of cusk, and 1,000 of hake worth $4,000, one of the highest stocks of the season. The fishermen earned $150 each for two weeks of exposure and discomfort. Parsons kept the *Pontiac*'s gross stock near the top in 1913 at $43,000; second in 1914 at $49,000; and first in 1915 at $53,735. If the *Pontiac*'s owners, represented by L. B. Goodspeed, received one-fourth of the gross stock as profit, the *Pontiac* more than paid for herself during these three years.[8]

A Variety of Fishing Craft in the South Dock, about 25 April 1913

THE SOUTH DOCK is crowded with some of the forty-five vessels that have arrived in the past three days. Now entering the dock is an Italian motorboat, with a man and several boys aboard. She has a Swampscott dory on deck and another towing astern for use in setting and hauling trawls.

Nearest the camera is a salt-stained schooner. Judging from her chain plate arrangement, she is a product of the James & Tarr yard.[9] The *Ripple* and another steam trawler lie along T Wharf;

Variety of fishing craft

Gyda

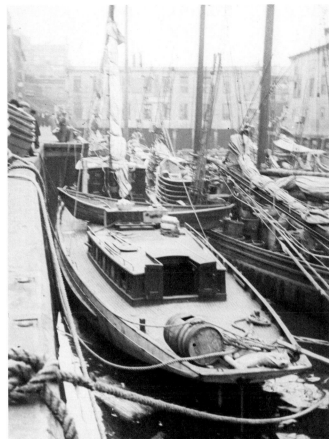

Gyda

outboard is the first knockabout fishing schooner, *Helen B. Thomas,* built in 1902. In the center is the powerboat *Mary F. Ruth,* seen under construction in the photographs of the Story yard.

In 1912, A. D. Story built three similar power-boats, the *Lois H. Corkum, Joanna,* and *Mary F. Ruth,* and another, the *Thelma,* in 1913. Among the earliest schooners with the helm forward, they looked quite similar to the diesel-powered drag-gers of the 1920s, and the so-called "western draggers" of the Stonington, Connecticut, fleet of the late 1930s and early '40s. Used for mackerel seining and gillnetting, as well as swordfishing, their practicality was proven in 1915 when the *Lois H. Corkum* was highline of the mackerel fleet, stocking $33,200. These twin-screw, gasoline-powered boats also foreshadowed the swift sur-plus submarine-chasers purchased for mackereling after the First World War.[10]

The *Mary F. Ruth* fished out of Gloucester until being sold to New Bedford in 1923. In 1927 she went to New York, and was lost in August 1928 at Isaac Harbor, Nova Scotia, while swordfishing.

The Former Yacht *Gyda* in the South Basin, about 25 April 1913

ALTHOUGH she too is a fishing vessel, the *Gyda* clearly shows her yacht heritage among the crowd of dory trawlers in the south basin. No doubt her quarters are luxurious, even compared to the usually finely finished fishing schooner forecastles and cabins. With a gas engine, her original schoo-ner rig, modified to sloop rig, is primarily for steadying. A fisherman's discarded mattress lies in the water by the *Gyda*'s stern.

Several fishing vessel designs were used for yachts and, conversely, a number of yachts were converted into fishing vessels. The most striking example was the transformation of the steam auxiliary brig yacht *Aloha* into the steam otter trawler *Heroine* in 1912. In Boston, the development of the motorized boat fisheries just before 1910 increased the demand for suitable powered craft. Trawling and, beginning in 1909, the Great Lakes form of gillnetting, made good use of old and new motorboats.[11]

Late in 1910, two ninety-five-foot Lawley steam yachts, *Bethulia* and *Philomena*, built in 1893, were purchased for fishing, as was *Geisha*, an eighty-five-footer built at New York in 1897. By the following fall, eighteen cod gill-netters, including *Bethulia* and *Geisha* of Boston, were fishing out of Gloucester.[12]

In 1913, Frank R. Neal & Company, commission fish dealers on T Wharf specializing in bait, purchased the former yachts *Wissoe* and *Gyda* for shore fishing. The *Wissoe*, a seventy-five-foot steam yacht built at New York in 1897, could rush herring, or other shore-caught fish, to market at ten knots. She was valued at $11,000.[13]

The sixty-five-foot *Gyda* served as the yacht of J. Hopkins Smith of Portland, Maine, for twenty years after her launch in 1893. She was a typically elegant example of the work of George F. Lawley & Sons of South Boston and Neponset, yacht builders from the 1870s through the 1920s. Frank R. Neal removed her compound steam engine and installed a gas engine when he fitted out the *Gyda* in 1913. She was used shore fishing by Neal until 1925, and then, with an oil engine, she fished for another four years as a small dragger. She foundered on Scituate Rocks, 20 November 1930.

Whaling Schooners at Merrill's Wharf, New Bedford, Winter 1916–17

THE FORMER fishing schooners *Margarett*, *Valkyria*, and *A. M. Nicholson* lie near Merrill's Wharf, one of the last whaling wharves in New Bedford. A few whale oil casks are visible on the wharf.

These schooners characterize the whaling in-

dustry in the early twentieth century. By 1885, economical schooners had become almost as common as the aging barks in the Atlantic whale fishery. One-third of the whalers clearing New Bedford that year were schooners, and the proportion increased thereafter. The short voyages of Provincetown whalers, sometimes called "plum-pudding" voyages, were generally pursued in schooners also.

The late whaling schooners were often retired fishing schooners, purchased cheaply to seek the meager profits that still existed in the industry. Chapelle noted that these schooners were generally sharper-ended than the schooners built for whaling. Some continued to sail into the 1920s.[14]

In March 1908, the *Fishing Gazette* reported overtures by New Bedford agents seeking fishing schooners to be modified for whaling, three or four having been purchased for that purpose in the previous two years. This was seen as a compliment to fishing schooners, as whalers had to be exceptionally strong and well-built.[15]

On the right is the *Margarett*, built at Essex by James & Tarr in 1889. She fished for nineteen years out of Gloucester until being purchased for whaling by G. R. Harris of Norwich, Connecticut, in 1908. In 1910, as she lay in Hampton Roads, ten of her crew initiated a work stoppage, refusing to sail until their demands for better food and payment of wages were met. The conditions aboard some of these small whalers near the end of the industry are suggested by Walter Hammond in *Mutiny on the Pedro Varela*. After three voyages from Norwich—the last whaling voyages from the New London area—*Margarett* was purchased by James Avery and berthed in New Bedford. Between 1911 and 1924 she made eleven voyages in search of sperm whales in the Atlantic. Her final cargo included 300 barrels of oil and sixteen aliens from the Cape Verde Islands.[16]

Bow-to is the *Valkyria*, built at Boothbay, Maine, in 1889. She fished out of Gloucester until she joined the whaling fleet at New Bedford in 1909 under John A. Cook. By the time she was retired in 1921, she had made eleven voyages.

At the end of the wharf, beyond the oil casks, lies another whaling schooner. Records indicate the *A. M. Nicholson* was in port at the same time as the *Margarett* and *Valkyria*. The *Nicholson* was

Whaling schooners

built by A. D. Story at Essex in 1900. Under the renowned Captain Solomon Jacobs, she landed the first mackerel of the 1908 season at Fortress Monroe, Virginia, 30 March 1908. Captain Jacobs sold her for a whaler at the end of 1908. She made twelve whaling voyages and, in 1923, became a "Brava packet," trading and carrying passengers between New Bedford and the Cape Verde Islands.[17]

Note the catboats, including a cat yawl, moored between the schooners. Before the advent of the diesel engine, most New Bedford fishing vessels were catboats or slightly larger sloops, some of them with auxiliary gasoline engines.

7 Steam Trawling

THE European model steam trawler became an increasingly common sight at T Wharf after 1910. Trawling—dragging a conical net bag across the ocean bottom—became the primary British method of bottom fishing before 1860. American experiments in 1864, and in 1891 with the British type ketch *Resolute,* suggested the feasibility of the trawl net in New England, but were not productive enough to challenge the hook and line methods in use. Only in Cape Cod's winter flounder fishery, followed in small sloops and schooners after 1895, did the beam trawl find consistent use in New England.

In the 1880s the British beam trawl fishery began to use the steam engine, and very quickly the steam trawler became the standard vessel fishing the Dogger Bank. Captain Joseph W. Collins, the greatest voice for reform in the American fisheries, called for the application of steam power about the same time, but only the oyster and menhaden fisheries, both pursued close to shore, adopted steam power to any degree before 1900. Exceptions were the steam mackerelers *Novelty* of 1885 and *Alice M. Jacobs* of 1902. About 1900, otter boards replaced the unwieldy beams that held open British trawls while fishing, and the developments soon to be transplanted to New England were complete.

It was not until 1904 that a serious effort was made to try steam trawling in New England. That year a group of investors, largely T Wharf fish dealers and Boston bankers, sent to England for plans of the latest model vessel, purchased rights to the patented British otter trawl, imported a trawl and other essential gear, and sent Captain H. Dexter Malone to observe on British trawlers.

The Quincy Fore River Shipbuilding Company built a steel vessel for the investors, now organized as the Bay State Fishing Company. Launched late in 1905 and christened *Spray,* the new steam trawler made her first set off Chatham on 19 December 1905. After an initial shakedown period, and after the crew was thoroughly trained by Fishing-Captain Hool, who was brought from England to assist Captain Malone in operating the *Spray,* the owners decided that the vessel was a success.[1]

However, Captain Malone became disenchanted with steam trawling after six months and returned to the schooner fleet. No doubt dory fishermen were heartened by his analysis.

A schooner of the fast sailing type which is now characteristic of the fishing fleet can beat the *Spray* in from South Channel by four hours, with a fair wind. A schooner and its dories can cover more ground than can be covered by the steam trawler. . . . I am convinced that month for month a schooner can bring in more fish than a steam trawler and the condition of fish brought in by a net are not to be compared with those landed by dorymen on a trawl.

Yet, because of the steam trawler's ability to fish in virtually any weather, Captain Malone saw a positive side to otter trawling.

Steam trawling is a much more humane method of fishing, and from that point of view deserves to be a successor of the dory style of trawling. From any observation, covering all points of trawl-

Ripple

ing in this country and England, I am convinced that it will be many years, if ever, before the schooner trawlers will be succeeded by steam.[2]

Despite its expense, its complications, claims that it destroyed fish stocks, and opposition among dory fishermen who saw their way of life threatened, the steam trawler was accepted more quickly than Captain Malone expected. The number of Bay State Fishing Company trawlers increased from one in 1905 to three in 1910, six in 1911, nine in 1913, and twelve in 1915. Other owners outfitted otter trawlers by mid-1912, and the wartime boom increased the number to fifty-five by 1920. Gradually, through the 1920s and '30s, the diesel otter trawler replaced the more expensive steam trawler and the schooner, and remains the standard New England fishing vessel today.

Steam Trawler *Ripple* at T Wharf, 16 February 1911

THE NEW steam trawler *Ripple* takes aboard ice on a warm winter afternoon. Up forward, an iced-up schooner, probably the *Elsie*, attracts a crowd's attention. One of the fish dressers, in his white coat, has stepped out of his work room and gazes down the wharf, while a few spectators stroll by purposefully or smoke their pipes in contemplation.

The *Ripple* has been in service for about a month and a half, and in that time has landed six trips of fish. Yesterday, she came in to deliver 60,000 pounds of haddock and 2,500 of cod to an active market with good prices. Through April she will continue the pace, averaging a delivery every 8½ days.[3]

The *Ripple* was the third steam trawler built by the Bay State Fishing Company. After 1906, a disastrous first year of experimenting with the *Spray*, training proper crews, and learning the bottoms where fishing was good, the company began to turn a small profit yearly, except during the recession period of 1908–09. In 1910 the company began to direct all profits into vessel construction, launching two trawlers in 1910, three in 1911, and three in 1913.[4]

By 1913, the nine Bay State Fishing Company trawlers delivered 16 percent of Boston's fish. They made 326 trips that year, averaging 36 each, or a turnaround time of ten days, a bit slower than the banner year of 1912 when the six active trawlers averaged 49 trips each and almost 51,000 pounds of fish per trip. In 1914, 72 percent of the otter trawling took place in the South Channel, with 17 percent on Georges Bank, and 11 percent on Western Bank. The 16,900,000 pounds landed comprised 18 percent of the fish delivered to Boston. Otter trawlers made up 18 percent of the net tonnage of Boston's fishing fleet, though several hundred other vessels also landed fish in Boston. Haddock was the most common species caught, totaling 26 percent of Boston's haddock and 65 percent of the haddock scrod (one to 2½ pounds in weight). The otter trawlers also landed cod, cusk, hake, and lemon sole.[5]

Steel trawlers like the *Ripple*, built by the Quincy Fore River Shipbuilding Company, cost between $50,000 and $70,000 ready to fish, two to three times the cost of a large schooner. Their insulated fish holds had a capacity of about one hundred tons of fish, as did large schooners. Still, it was estimated that a steam trawler could land $50,000 worth of fish a year, given luck and high prices, and the breed proliferated.[6]

Ripple in the South Dock, about 25 April 1913

THE *Ripple, Mary F. Ruth,* and several schooners are in. The *Ripple* came in on the twenty-fourth with 50,000 pounds of haddock and 3,500 of cod.

Some of the details of a steam trawler, including a trawl, trawl door, riding sail, and the entrance to the officers' and engineers' quarters are visible in this view. It appears that a couple of fish are drying on the engine room vent beside the open companionway. On top of the trawl is a tub that apparently contains the deep sea lead. Like the schooners, the trawlers used the lead to sound the ocean depths.

The trawl is composed of various weights of manila twine, woven into a conical net bag, 110 feet across and 4 feet high at the mouth, tapering back 130 feet to the small "cod end." A door like the one along the port rail is attached at either end of the mouth to swim outward and keep the trawl open as it is dragged across the bottom at three miles-an-hour for about an hour and a half at a set.

Doors were more convenient than the large beams used by the English sailing trawlers and the early steam trawlers. Barring rocky bottom or very bad weather, a trawler would always have one otter trawl towing on the bottom and the fish from the other trawl on deck being dressed down while any necessary repairs were made to the second trawl.

Ripple in the south dock

Not only was the price of a steam trawler several times greater than that of a schooner, but the trawler's gear was a great deal more expensive as well. An otter trawl cost about $3,000, and most trawlers used two alternately, for a total of $6,000 in gear, or ten times the value of the longline trawls carried by a schooner. The steam trawlers were restricted to sandy bottom, especially the South Channel and parts of Georges Bank, but still the otter trawls wore quickly and required frequent repairs.[7]

Steam Trawler *Spray* Taking on Ice, T Wharf, Fall 1913 and 19 February 1914

IN THE first view, bystanders look on as fishermen, aided by the deck hoister, load the last few blocks of ice aboard. The vessel must be the *Spray*, first of the Bay State Fishing Company trawlers, as she was the only one without a turtleback topgallant forecastle deck. To enter the *Spray*'s forecastle hatch (obscured by the gallows and the fisherman standing on deck in this photograph) on the flat, raised forecastle deck must have been a wet ordeal in a head sea. The turtleback deck, with protective bulkhead shielding the forecastle hatch in the after end, was a great improvement in later trawlers.

In the second view, the *Spray* is loaded with ice after delivering 25,000 pounds of haddock and 800 of cod. The unprotected foredeck and forecastle hatch can be seen to the right of the mast. Her ice coating is evidence of the heavy weather of early 1914. In the background are some of the eleven fishing schooners that also arrived on 19 February, and the coaster *City of Augusta*.[8]

The Bay State Fishing Company trawlers carried about ten tons of ice for preservation. Some had crushers on deck to feed ice into the hold, whereas schooner fishermen chipped their ice by hand. The engine of the ice crusher served as the deck hoister as well.[9]

Built in 1905 on the latest British model by the Fore River Shipbuilding Company at Quincy, the *Spray* was the Bay State Fishing Company's only vessel until 1910, during which time she proved the efficacy of offshore otter trawling in New

Spray

England waters. She continued to provide good service until 1916, when the company replaced several of its trawlers. Then, after about three years of ownership in Boston, she was sold to the Russian government. Whether or not she crossed the Atlantic to fish from a Russian port, by 1922 she was serving as a tugboat out of New York City for Anthony J. McAllister. The *Spray* returned to Boston's fishing fleet after being sold to the O'Hara Brothers in 1928. Renamed the *Patrick J. O'Hara,* with a diesel engine, she fished until being broken up about 1943.

The Bay State Fishing Company remained in business until about 1960. After a management change in the late teens to give the firm a better business footing, it became known for its actively marketed "40 fathom fish." The company claimed credit for conceiving of, and naming, the fish fillet, about 1920. The Bay State fleet grew to nine vessels by 1914, fourteen by 1920, seventeen by 1929, and a high of nineteen in 1932. Its first diesel trawlers, named *Ebb* and *Flow,* were built by Bath Iron Works in 1929. Between 1938 and 1940, most of the fleet was sold off, and for its last twenty years the company operated only one vessel.[10]

Steam Trawlers at the Coal Dock, Fall 1913

U P the Chelsea River, near the Meridian Street bridge, a steam trawler, right, refills her bunkers at the Metropolitan Coal Company wharf while another trawler maneuvers in the channel. In the center, barge number nine of the Lehigh & Wilkes-Barre Coal Company discharges anthracite from northeastern Pennsylvania in the coal pocket.

Steam trawlers were dependent on Pennsylvania coal miners in order to operate. They carried bunkers sufficient for fifteen or sixteen days of continuous steaming, though most trips were half that long. Fitted with efficient triple expansion engines and Scotch (firetube) boilers, they developed 450 indicated-horsepower at 110 revolutions per minute, with a maximum speed of about ten knots.[11]

The trawler in the stream appears to be the *Breaker,* one of three new vessels added to the Bay State Fishing Company fleet in 1913. Launched in August, only a few months before this photograph was taken, her boat covers and triced-up jib are still white and fresh. The *Breaker* put in a grueling but relatively uneventful twenty-three-year career before being abandoned in 1936 to be replaced by a more economical diesel-powered vessel.

Steam Trawler Getting Out a Trip of Fish, Fall 1912

S OME of the crew are hard at work, while others have already changed into their shore clothes. Forward, the man in shirt sleeves operates the combination steam donkey engine and ice crusher (ice was loaded through the hatch propped open just below the mast). A revolving bollard of the steam hoister used to haul back the trawl, similar

Coal dock

to those in use on steam oyster dredges, is visible under the corner of the bridge deck. Pen boards can be set up between the square stanchions on deck to form "ponds" for sorting the fish when they are dumped from the otter trawl. The basket emerging from the hold appears to be filled with sole.

Up on the protective turtleback topgallant forecastle is the capstan. Steam trawlers carried a minimal sailing rig, primarily for steadying, and the triced up jib is visible. Abaft the mast is "Charlie Noble," the galley stove pipe. As in the schooners, most of the crewmen live forward; the forecastle has pipe berths for fourteen men.

The first American steam trawler, *Spray*, had a mate and two fishermen from the English fleet to train the novice crew.[12] Steam trawler crews, including captains, averaged about seventeen men. Captain Malone complained that steam trawlers were at a disadvantage to schooners in crew size. A steam trawler did not carry enough men to dress down the fish conveniently when the three-ton haul of the trawl was dropped on deck, and yet did not provide enough work to occupy a larger crew at other times.

The division of labor aboard a steam trawler made her operation very different from that of a schooner. Trawlers followed a six-hour watch system around the clock. At any time, either the captain or mate, an engineer, fireman, and perhaps six fishermen would be on duty. The steward or cook worked equally as hard on a steam trawler as on a schooner.[13]

Steam trawling required both skilled fishermen and skilled engine-room operatives. Unlike the men of the schooner fleet, steam trawler crews were paid a wage rather than a share of the profits. The wage system opened the way for labor actions, first among the specially skilled and hard to replace engine-room crews, who were members of the large Marine Fireman's Union. The seamen and firemen went on strike in July 1912, demanding better pay and shorter hours. After tying up four of the five trawlers for a week, they settled for an increase from $10 a week to $50 a month. Again, at the end of 1915, steam trawlermen struck, this time to gain recognition of their union. The *Breaker* got to sea with a crew of longshoremen, and there was talk of strike-breaking crews

Getting out a trip of fish

from New York, but it was clear that an untrained crew was virtually useless. It was not until 1917 that schooner fishermen called a general strike in New England.[14]

An Icebound Steam Trawler Discharging Fish, about 19 February 1914

A LONGSHOREMAN and a well-dressed spectator or fish dealer survey the ice-encrusted vessel while the fishermen concentrate on filling fish boxes like the one on the scale in the foreground. The vessel is probably the *Surf*, just arrived with 38,000 pounds of haddock and 1,000 of cod. Note the ice-sheathed electric lights mounted on the bridge rail to illuminate the deck for night work.[15]

The weather in January and February 1914 was the worst in years, and icebound vessels were commonly seen at T Wharf. The steam trawler *Breaker* arrived on 12 February sheathed in a foot of ice after spending a night off Highland Light in a northwest gale.[16]

Obviously, winter fishing aboard a steam trawler was not easy, but it was somewhat less dangerous and uncomfortable than was winter dory trawling. The helmsman aboard a steam trawler even had a heated wheelhouse! Nor did the steam trawlers lose a man until, in 1912, the converted yacht *Heroine* lost a fisherman overboard on Georges Bank, and the *Spray* lost a man killed by gear falling from aloft.

By the summer of 1912, the increase in the number of American steam trawlers galvanized the previously sporadic and isolated opposition to otter trawling. The most damaging charge claimed that the otter trawl was overly destructive to fish stocks and spawning grounds. Admittedly, the otter trawl swept up fish indiscriminately; a 1914 study determined that only half the fish were marketable in summer. But its supporters claimed that "there is not the slightest danger that man can reduce the supply of fish in the ocean."[17]

Opponents pointed to the depletion of New England river fish as evidence that man can indeed reduce fish stocks, and to evidence from British otter trawl fisheries for examples of the damage caused by otter trawls. Scientific evidence was contradictory, cited by both sides, or completely lacking.

Dory fishermen felt that the otter trawl threatened their occupation and their way of life, which many felt was well worth saving. The Massachusetts Fisheries and Game Commission biologist who studied otter trawling betrayed his bias when he suggested, "hand lining is thoroughly American in the sense that it comes nearest to giving equal opportunities to all, and under natural conditions is adequate for furnishing a reasonable supply of food."[18] Despite claims that otter trawlers and other fishermen coexisted in Britain, American fishermen feared that the trawlers would drive the schooners from the seas.[19] Reports emphasized their concern at losing their independence. Captain Frank A. Nunan presented the scenario:

It is labor and independence against something not right. By way of an illustration, take the day before Christmas, or the day before a big storm.

The Old Way—'Well boys, go home. We won't go out tonight'.

The New Way—'Well, wife, the fish company says I must go out tonight. You know I am hired and I must go, for if I don't there is someone to take my place on the wharf. Goodbye. Think of me tomorrow'.[20]

They feared that "the crews of the otter trawlers do not need to be fishermen. . . . They can come from Mattawamkeag as well as from Orr's Island, and it matters little whether they can splice a rope or make a Turk's Head or whether their knowledge of rigging is confined to tying a knot around a cow's neck so it will not slip and choke her to death."[21]

Finally, it was claimed by some that steam trawling was yet another sinister manifestation of the rise of business trusts. Looking at the short-lived National Fisheries Company of 1906, which had sought to combine the Boston fish dealers, control the price of fish, and promote steam trawling in New England, the wary saw the threat of a similar business venture. The fact that the Bay State Fishing Company had been included in the National Fisheries Company made them all the more concerned about its growth.[22]

The demand for legislative action against steam trawling brought forth serious scientific investigation. In March 1912, the Massachusetts Fish and Game Commission initiated a study of the effects of otter trawling, which was used as evidence before Congress and published in 1915. Statistics, collected on a single trip, suggested that 60 percent of the fish taken were marketable, 21 percent were either undersize or otherwise wasted, and 19 percent were unmarketable species. The report identified no other blatant ill effects but recommended that otter trawling be thoroughly studied in a restricted environment. The U.S. Bureau of Fisheries report, published in 1914, was broader based, finding in summer the catch of unmarketable fish, including immature fish, was significant—as much as 60 percent. Otherwise, it produced similarly ambiguous evidence and also proposed restricting the fishery pending thorough study. Consequently, in 1915, otter trawling was limited to Georges Bank, the South

Icebound steam trawler

Channel, and Nantucket Shoals, the areas where the great majority of otter trawling occurred anyway.[23]

The diesel engine democratized otter trawling in the early 1920s, making small, individually owned trawlers feasible. The center of activity moved inshore (where sail and gas-powered otter trawlers had quietly been hunting flounder for years), and the controversy died down. Despite the rhetoric, if some of the skills of dory fishermen were no longer required, there was still a need for fishermen adept at handling fish and familiar with the ways of the sea. Now, of the three bases of opposition to otter trawling, only the claim that it is destructive to fish stocks remains in contention.

Launched in December 1911, the *Surf* was the fifth Bay State Fishing Company trawler. The company began to equip its vessels with radio transmitters about 1925, but the *Surf* did not receive one until 1930. She was abandoned in 1936, after twenty-three years of routine service.

Ripple

Steam Trawler *Ripple* in the North Dock and *Crest* in the South Dock, Winter 1911–12

THE *Ripple* lies between two schooners with one of the tireless T Wharf waterboats, originally rigged as a catboat but now powered by gasoline, alongside. On the other side of the busy wharf the *Crest,* a Quarantine Department steamer (right), and several schooners lie in the quiet basin. The last office in the T Wharf block houses the Ross Towboat Company, and just to the left of that is the entrance to the New England Fish Exchange auction room.

The *Ripple* was launched by the Fore River Shipbuilding Company at Quincy in the fall of 1910, about six months after her sister *Foam.* Under Captain Michael Green she landed her first trip early in January 1911. Like the market schooners, *Ripple* and her sisters averaged about a trip a week. This is where they made money: consistent trips and quick turnarounds.[24]

On 13 July 1911 the *Ripple* was involved in an accident which pointed up one of the hazards of otter trawling. In a dense fog sixty-five miles southeast of Highland Light, *Ripple* was cruising slowly, sounding her whistle at intervals. When a faint horn was heard through the fog, the skipper reduced speed, but suddenly a schooner appeared dead ahead. With engine reversed, the *Ripple* plowed into the fishing schooner *Independence II.* The schooner did not sink, but was damaged bad-

Crest

ly enough that she was towed back to Gloucester by the *Ripple*.[25] Such collisions might have been more common had the steam trawlers visited all the grounds frequented by the schooners. As it was, only a few areas provided the consistently smooth bottom sought by the steam trawlers.

In 1912, the ubiquitous Captain Solomon Jacobs, the great mackerel catcher and innovative skipper who had gone otter trawling on Ipswich Bay in a gasoline-powered boat before 1905, was injured by the wheel while in command of the *Ripple*.[26]

The Bay State Fishing Company sold the *Ripple* in 1917, replacing her with a new, larger *Ripple*. The original continued to fish out of Boston under a new owner, then was sold to New York. Renamed *Boston*, the *Ripple* returned to Boston to fish for the Massachusetts Trawling Company in 1927. Reequipped with a 330-horse-power diesel engine in 1929, she put in another nineteen years, until being dismantled in 1948.

On 29 March 1911 the Bay State Fishing Company officers traveled to the Fore River Shipbuilding Company in Quincy to launch their fourth steam trawler, the *Crest*. The eight-year-old daughter of Harrison I. Cole, manager of the company, did the honors as the *Crest* slid down the ways. After twenty-five years of service, the Bay State Fishing Company abandoned her as unfit in 1936.[27]

Otter Trawler *Long Island* at South Boston, about 7 August 1914

BEYOND a jumble of fish carts, one owned by the Atlas Fish Company, lies the *Long Island*, held off the pier by wooden "camels." Her origin is revealed by her hull design; she was built as a menhaden purse seiner. But the characteristic "gallows" above her starboard rail identify her as an otter trawler.

The *Long Island* was built at Rockland, Maine, in 1912 for the Atlantic Fertilizer & Oil Company of Greenport, Long Island. She continued as a menhaden steamer through 1913. But when her sisters were dividing their time between menhaden and tinker mackerel in 1914, the *Long Island* was proving herself as an otter trawler.

The *Long Island* and the *Heroine,* a large steam yacht converted to otter trawling in 1912, demonstrated the feasibility of using vessels other than the specially constructed North Sea model built by the Bay State Fishing Company. The Trident Fisheries Company of Portland fitted out the *Long Island* and she began to land fish at Boston in April 1914. During the busy month of August 1914 she landed fish on an average of every five days, for a total of 260,000 pounds of haddock, 13,400 of cod, 1,000 of pollock, and 100 of hake. In June 1915 she landed the two largest fares taken by an American otter trawler to that time: 280,000 and 300,000 pounds.[28]

Although she had a more shoal draft hull and carried a larger crew than most otter trawlers (twenty-five as opposed to seventeen) she apparently proved to be economically successful. As a used wooden vessel, her cost must have been significantly less than that of a steel otter trawler, even considering the necessary transformation of the fish hold and addition of trawling gear.

Long Island

Seal

Toward the end of the wartime fish boom, the *Long Island* was apparently replaced, perhaps by one of the new wooden otter trawlers. After disappearing from the records during 1918 and 1919, she turned up in Reedville, Virginia, back in the menhaden fishery. Fifteen years later, in 1935, she was sold to Wilmington, Delaware. She foundered 18 September 1936 in heavy seas on Delaware Bay, with the loss of seven of her crew of forty-two.[29]

Steam Trawler *Seal* Almost Ready for Launch, Story Yard, Essex, Summer 1917

THE WOODEN steam trawler *Seal* lies on the launching ways, with her rudder braced and her rudimentary masts rigged. The sheet-metal guards on her hull will protect the planking from the heavy, metal-shod trawl doors. Behind the *Seal*, the schooner *Hesperus* is being planked. Forward of the *Seal*'s pilothouse, a ginpole is rigged up by the stem of the fish freighter *Gaspe*.[30]

Once the Bay State Fishing Company's steel otter trawlers had proven themselves, other fishing companies took up otter trawling. Due to the price of the steel trawlers, they used a variety of new and second hand wooden vessels. When the venerable Benjamin A. Smith Company determined to bring otter trawling to Gloucester during the fish boom of the First World War, they went to Story and James at Essex for their vessels. Theirs were the first two otter trawlers built at Essex, the *Walrus* by James and the *Seal* by Story. They also hired two of the best skippers out of Gloucester, Captain Clayton Morrissey for the *Walrus* and Captain Lemuel Spinney for the *Seal*.

Powerful vessels, with 650-indicated-horse-power triple expansion steam engines and large crews of twenty-four men, they could turn a profit during the boom years between 1917 and 1920, but were too expensive to operate as the demand for fish declined during the postwar economic contraction. Few large otter trawlers were built in the early twenties, and those that were had oil engines rather than steam. Schooners continued to be built, and there was a surge in the construction of small wooden diesel-powered otter trawlers to fish inshore.[31]

The *Seal* was sold British about 1924, then to Joseph O'Boyle of New York in 1926. Sold to the Hochelaga Shipping and Towing Company of Halifax in 1928, she was renamed *Guard*. Serving as a salvage vessel, she was lost at Seal Island, Nova Scotia, that year.

8 Italian Boat Fishermen

ITALIAN FISHERMEN did not become a force in the Boston fisheries until about 1900, as Boston's Italian population grew by immigration, particularly from Sicily. Beginning with Swampscott dories, they rowed and sailed many miles up and down the harbor each day, as had Yankee fishermen in the nineteenth century and Irish fishermen after 1840. By 1908, the shift to gasoline-powered boats in the Italian fleet was well along.

The Italians caught flounder and other fish with trawls, herring with torch and dip net until the method was outlawed in 1911, a variety of fish with gill nets and pound or trap nets, as well as eels, crabs, and other shellfish. The large boats supplied Italian wholesale firms that shipped fish to Italian colonies throughout the country. Fishermen owning smaller boats sold their fish at retail on T Wharf or in the streets of the North End. Because they were major suppliers of the poor, they were allowed to sell in the open, despite health laws and other prohibitory ordinances.[1]

The Italian fishermen remained at T Wharf and the Eastern Packet Pier after the fish dealers moved to South Boston in 1914, managing themselves and increasing in importance. Part of the fleet moved to Gloucester shortly after 1910, becoming a significant part of the fresh fishery there. To this day Italian fishermen are a dominant force in the Boston and Gloucester fisheries, and are constantly reinforced by immigration.

An Italian Fishing Boat Comes in to T Wharf, about 19 February 1914

THE ICY BOWSPRIT of a market schooner overhangs an Italian power dory with a load of fish aboard. A crowd of spectators size up the catch.

In April 1908 the *Fishing Gazette* reported,

The Italian fishermen at 'T' Wharf, Boston are becoming more of a factor daily. During the week they were successful in bringing quantities of fish to market in

Italian fishing boat

their little boats. They swarm the docks like flies, live on bread and water until they get money enough to buy a gasoline engine for their boat, after which they inaugurate a regular business, and supply their customers daily with stock, principally flat fish.[2]

Italian immigration into the United States was relatively late, being minimal until 1880, and it was not until the decade 1900–1910 that it became significant. Poor conditions in the south of Italy in the 1890s contributed to the increased exodus of Italians to America. They grew from 5 percent to 10 percent of the foreign-born population between 1900 and 1910. While less than 15 percent of the Italians settled in New England, Boston had a sizable population—31,000 in 1910 and 43,000 five years later.[3]

In southern Italy, where the majority of Italian immigrants to the United States originated, fishing was a common occupation along the coast, and Italians made up a large percentage of the European fishermen who immigrated. Italian fishermen were common in San Francisco and several fishing communities in the South by 1880, but did not appear in any numbers at Boston until about 1900 (only 3 of 14,676 Massachusetts fishermen in 1885 were Italian-born). Soon they had a sizable fleet of Swampscott dories with sailing rigs to fish Boston Bay and the nearby shore. Within four years many of them had embraced the gasoline engine to pursue their work.[4]

One story has it that in 1904 two Italian fishermen, tired of rowing or sailing their Swampscott dory fifteen or more miles a day, noticed a powered yacht tender speed by on its way up the harbor. They installed a four-horsepower motor in their boat, and once they learned the intricacies of it, the benefits were obvious.

When the first motor fisherman got well into the harness her owners' wives began to appear on the street in new dresses and wore smiles of contentment while promenading the North End of Boston, where the Sicilians have their homes. The boat was a wonder in the way of earning money. She made two trips, sometimes three, while the sailboats were getting their fares to market. A bigger success on a small scale never appeared in the harbor.[5]

By 1908, most of the Sicilians were using power dories like this one.

The Sicilian Fishing Boat Fleet at the Head of the South Dock, 1913–14

A VARIETY of designs are visible. On the right is one of the largest of the Sicilian boats, with a standing gaff rig and two Swampscott dories on deck. A boat of this size acted as a mother ship for several dory fishermen. A horseshoe is tacked to the bulkhead for luck.

Slightly smaller, and more numerous, are the carvel planked craft with high, almost plumb, stems, and two heavy rubbing strakes. This model, which apparently began to appear about 1908, bears a resemblance to the Irish fishing cutters which had been in use until shortly after 1900 in Boston harbor. The one in the center has a shallow tub of flounder trawl on the cuddy top.

The small motorboat on the right represents the earliest type of Italian motorboat. It was modified from the lapstrake Swampscott dory, lengthened to about thirty feet, with a deepened bow and more bearing aft to accommodate the engine's weight. The twenty-five- to sixty-gallon gasoline tank was located in the very bow. The boat on the right illustrates the division of the cockpit into a standing room for hauling trawls and a fish hold amidships.

The small boats had six-horsepower, single cylinder engines; the larger boats might carry twelve-horsepower, two cylinder motors, and the very largest had twenty-one-horsepower, three cylinder power plants, giving them a top speed of seven or eight miles an hour. In the cuddy aft were the engine, wheel, knife switch starter, and narrow seats on which the crew could sit, or even sleep, in bad weather. The small boats took a crew of two; the others three or more.

The small boats were built at Swampscott or East Boston for about $250, without engine, in 1908. They lasted for five or six years of almost constant use, operating eight or more hours virtually every day. Dockage charges were about fifty cents a week, and gas, purchased from gas boats, was ten cents a gallon in 1908, rising thereafter. Expenses were limited to gas, oil, batteries, and occasional repairs to bent propellor shafts. With such limited maintenance, the simple Mianus make-and-break engine was the ideal power plant

Sicilian fishing boat fleet

for this type of vessel. All in all, they were safe, durable, efficient, productive craft.[6]

The schooner on the right may be the big *Regina,* which was laid up at Long Wharf from late 1911 to late 1913.

Italian Fishermen in the South Dock, late April 1913

THREE FISHERMEN straighten up their boats as their compatriots sell their catch along the stringpiece. The boats are Swampscott dories, used by the Italians before they adopted powerboats, and power craft that bear a resemblance to the "Boston fishing cutter" hulls introduced by the Irish boat fishermen who preceded the Italians.[7] The small tubs hold flounder trawls, which were often baited by the fishermen's families at home.

In the background is the brick building at the inner end of T Wharf, built about 1875. The bay window announces the Master Mariners' Towboat Company.

The fishermen are straightening up their boats, but they are not cleaning fish at T Wharf. A 1908 ordinance prohibited fish cleaning near the wharf because decaying refuse in the water was a health hazard and made T Wharf a noxious place to visit.[8]

All fish commonly caught alongshore were

Italian fishermen

found displayed in the Italian fishermen's baskets on T Wharf. Flounder was caught in the warmer months; cod, haddock, and hake were taken in winter; and in fall herring were caught by torching at night. The light of a torch on a pole attracted herring to the surface where they could be taken with a dip net, in centuries-old style. When this form of fishing was prohibited in 1911, some Italian fishermen began to use gill nets. They also caught a variety of fish in traps, weirs, or pound nets.[9]

In 1912, Italian fishermen taught Boston consumers a lesson. They suggested that albacore—horse mackerel or what we call tuna—was a good substitute for high-priced beef.[10] Before this time, albacore had not been a popular fish. In 1908, only forty-five tons, worth $5,400, were landed. All of it was brought in by boat fishermen tending traps or weirs.[11] The Italians also pioneered the use of other neglected species, including whiting, skate, shark, and dogfish.[12]

Italian Fishermen Selling Their Catch, T Wharf, Fall 1912

THE BOAT FISHERMEN paid a small fee to sell their catch on the south side of T Wharf. Almost any day a visitor could report:

I stopped to watch the Italian fishermen, with their baskets of flounders. These men squat in a row and invite, it seems, discussion with a buyer. "See! Fine! Justa right. You want him? Cheap! No! How much you give, mista?"[13]

This cry continued even after fish dealers abandoned T Wharf for modern quarters in South

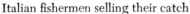

Italian fishermen selling their catch

Italian fishermen selling their catch

Boston in March 1914. On the new fish pier, boat fishermen were permitted to land fish for dealers, but they were prohibited from selling at retail on the pier. Since a good portion of their fish went to North End retailers, and the "small truck"—small fish—was sold in the open, as seen here, the Italian boat fishermen and some of the Portuguese maintained their lucrative business on T Wharf and the Eastern Packet Pier.[14]

When some of the Boston skippers noticed that T Wharf continued to function profitably, they revived the Fishing Masters' Association with the intention of sponsoring the return of part of the haddock fleet to T Wharf. In February 1915 some of the stalls reopened. However, to ensure that the wharf did not again become an unsanitary public nuisance, open display of fish was prohibited. This legislation did away with informal sale of fish by individual fishermen.[15]

Crabber at T Wharf, late April 1913

ONE of the Sicilian boat fishermen unloads a good haul of crabs at a crab car in the south basin. His double-ended power dory has a steering yoke on the rudder. Atop the cuddy cabin are two crab pots, which are simply hoops with net bags attached.

Crabbing was one of the small seasonal boat fisheries pursued in Massachusetts. For example, in 1908, 123,000 pounds of crabs (mostly hard crabs) were taken in Massachusetts waters, 116,000 by Suffolk County fishermen like this one. They were worth $2,600. By contrast, Maryland, known for its crabs, produced almost 20,000,000 pounds, worth $225,000. Some were caught with dip nets, like the one on the crab car, but most were caught in baited crab pots like the ones on the boat.[16]

Crabber

APPENDIX I: Vessel Specifications[1]

Name	Tonnage Gross	Net	Length[2] (feet)	Breadth (feet)	Depth of hold (feet)	Built	
A. M. Nicholson	136	100	105	24.6	10.4	Essex	1900
Alaska	229	121	141.8	21	10	E. Boothbay	1881
Appomatox	69	47	78	21.4	8.6	Essex	1902
Arethusa	157	107	114	25.6	12.5	Essex	1907
Aspinet	126	83	102.5	23.7	12.2	Essex	1908
Athena	94	62	100.5	22.6	9.8	Gloucester	1908
Bay State	159	109	112.7	25.4	12.3	Gloucester	1912
Benj. F. Phillips	138	102	110.8	25.8	11.2	Essex	1901
Breaker	248	119	117.8	22.5	11.6	Quincy	1913
Clintonia	147	105	109	25.1	11.9	Gloucester	1907
Commonwealth	141	93	103	24	11.8	Essex	1913
Crest	244	113	114.3	22.5	11.6	Quincy	1911
Delphine Cabral	119	77	101.6	23	11.5	Essex	1912
Dixie	17	16	48.4	15.3	5.5	Beverly	1880
Dorothy II	70	44	79	21.3	8.6	Essex	1904
Eliza A. Benner	21	14	47.1	16.9	6.7	Waldoboro	1900
Elsie	137	98	106.5	25	11.5	Essex	1910
Ethel B. Penny	94	56	99.5	22.7	9.8	Gloucester	1908
Evelyn M. Thompson	97	57	104.4	22.7	9.8	Gloucester	1908
Finback	159	122	121	25.2	12.4	Essex	1916
Flora S. Nickerson	107	73	94.3	23.9	10.6	Essex	1902
Francis J. O'Hara, Jr.	117	83	104	24	11.2	Essex	1904
George Parker	133	100	97	25.6	10.6	Essex	1901
Gertrude DeCosta	105	61	102.5	23	10.6	Essex	1912
Gov. Foss	130	88	105.9	24.6	10.8	Essex	1911
Gyda	25	12	64.8	12.7	6.1	Boston	1892
Henrietta	99	62	89	22.1	10.9	Essex	1915
John Hays Hammond	132	92	101.9	24.5	11.8	Essex	1907
Josie and Phebe	137	88	107.9	25.8	12.2	Essex	1908
Knickerbocker	159	101	112.5	25.2	12.2	Essex	1912
Leonora Silveira	106	63	92.6	23.4	11	Essex	1912
Lillian	28	14	51.4	18.1	7.8	Gloucester	1902
Lillian (sloop)	12	6	34.1	13.1	6	Friendship	1910
Long Island	390	167	151.6	24.1	13	Rockland	1912
M. P. Howlett	85	55	84.8	23	8.8	Essex	1901
Mabelle E. Leavitt	21	19	47.8	15.2	7.1	Bristol	1900
Margarett	138	104	92.2	22.7	8.8	Essex	1889
Mary	140	93	113.8	25.7	12	Essex	1912

---

[1] based on U. S. Department of Commerce *Annual Lists of Merchant Vessels of the U. S.*
[2] measured between the forward edge of the planking at the stem and the after part of the rudderpost, at the level of the deck.

Name	Tonnage		Length (feet)	Breadth (feet)	Depth of hold (feet)	Built	
	Gross	Net					
Mary DeCosta	101	69	88.6	22.8	10.4	Gloucester	1909
Mary E. Silveira	93	63	85	22.4	10.2	Gloucester	1904
Mary F. Ruth	46	33	65	16.4	8.4	Essex	1912
Mary J. Beale	33	20	58.7	17.3	6.7	Eastport	1908
Motor	38	18	61.8	18.4	7.8	Gloucester	1903
Muriel	120	83	104.9	24.3	11.3	Essex	1904
Pontiac	115	79	96	23.7	11	Gloucester	1906
Reading	138	92	105.6	25	11.6	Essex	1914
Regina	147	111	115	25.7	11.6	Essex	1901
Ripple	244	114	114.3	22.5	11.6	Quincy	1910
Robert & Arthur	110	78	95.2	24	10.6	Essex	1902
Robert & Richard	140	89	108.8	24.8	11.1	Essex	1914
Rose Cabral	94	89	86.6	23.9	9.1	Essex	1890
Rough Rider	10	8	38.6	10.2	5	Manitowoc	1904
Ruth	90	56	85.4	22.1	10.8	Essex	1912
Ruth and Margaret	118	77	102.5	23	11.4	Essex	1914
Seal	474	249	163.1	26.2	13	Essex	1917
Shenandoah	110	69	86.4	24	9.2	Essex	1889
Smuggler	119	91	103.6	24.2	10.6	Essex	1902
Somerville	129	82	103.2	24.7	11.4	Essex	1914
Spray	283	159	126.6	22	12.9	Quincy	1905
Stranger	52	28	73.6	19.4	9	Essex	1903
Surf	252	119	117.8	22.5	11.6	Quincy	1911
Valkyria	139	104	92.6	24.1	10.2	E. Boothbay	1889
W. M. Goodspeed	94	64	100.1	22.7	9.8	Gloucester	1908
Walter P. Goulart	84	55	83	21.7	10.2	Essex	1904
Waltham	82	44	75.8	21.9	9.8	E. Boothbay	1904
Yucatan	84	52	93	22.3	11	Essex	1912

Atlantic & Pacific Fish Co. This company was organized in 1907 with Francis J. O'Hara as president, and Francis J. O'Hara, Jr., as treasurer, located on Atlantic Avenue. The elder Mr. O'Hara had been in the fish business for many years on Atlantic Avenue, had been very prosperous, and had built up a large and at that time a still growing trade. Shrewd and cautious, not given to display, but always ready to buy whenever a bargain offered, he was quite often in condition to sell goods at low, but at the same time, profitable prices. Mr. O'Hara, Jr., from the time of his first going to work, had always been employed with his father, but seeing the need of a larger store, and better accommodations, this new company was organized, and at once sprang into a goodly business of its own. Mr. O'Hara is in the very prime of life, going sanely and safely about his work, caters to both local and far-away trade. He came to the new pier with the others, and is located at No. 21, where from 7 to 5 a buyer always finds courteous treatment, and if the seller has a bargain to offer, he quite readily finds a buyer.

The Atlas Fish Co. The Atlas Fish Co. was organized on Oct. 31, 1907, and located at 78 Commerce street, Boston. Patrick J. Anderson, president, and Carlton H. Rich, treasurer, and associated with them were Capt. John Dench and Charles M. Parsons, a merchant from Gloucester. Mr. Anderson had been employed for 12 years by the B. F. Phillips Co. as foreman of their store on T Wharf, and Mr. Rich had been employed by the same firm as office manager for 16 years, while Capt. Dench had been master of a fishing schooner for many years, beginning life early as a fisherman. They started each with large experience both in the producing and the handling of fish. They designated themselves as "the house where quality counts." They are doing, and have done a general wholesale business and are the only fresh fish dealers in Boston, with one exception, who are members of the Oyster Growers and Dealers' Association of North America. They are both genial and courteous to customers and have built up a large and still increasing trade, dealing more with supplies for retail markets than with the larger houses throughout the country. After the first year Capt. Dench and

Mrs. Parsons retired from the business, selling their stock to Mr. Anderson and Mr. Rich, Capt. Dench going into the lobster business, in which he still remains. They remained on Commerce street until last April when, with the other wholesalers, they removed to the New Fish Pier, where they are located at No. 39.

J. Adams & Co. Across the sea in Sweden, the land of the Northern Lights, there was born in 1833, a boy who bore the name of Hans Peter Wennerberg, who was destined to play quite an important part in the fishing trade in New England. He appeared in this country at the early age of 18, as an officer on a ship. Somehow he became enamoured of America; somehow he drifted into Gloucester, where he early engaged in the halibut fisheries. He rapidly rose to be captain and became one of the famous halibut catchers of that famous old town, followed this business until 1869, when in 1870, together with Joseph Adams, they organized themselves into a partnership under the present firm name. They remained together until 1880, when Mr. Wennerberg purchased the Adams interest in the business, which he carried along alone until his death, which occurred in 1901. They first begun business on Commercial Wharf. In 1884 they removed to T Wharf, where after Mr. Wennerberg's death, the sons have carried on the business under the same name, until their removal last April to the new fish pier. The partners to-day are Henry P. Wennerberg, who is one of the trustees of the T Wharf Corporation and a director in the New England Fish Exchange, and his brother, James B. Wennerberg, both sons of the elder Mr. Wennerberg. The father of these boys was for many years treasurer of the New England Fish Co., a director of the T Wharf Co., and one of the stockholders of the Union Ice Co.

Arnold & Winsor Company In 1881 Orson Arnold and Sanford C. Winsor formed a partnership and went into the general wholesale fish business at 75 Commercial Wharf. Arnold had previously been with the firm of George C. Richards & Co. as foreman since 1879, and Winsor had been bookkeeper for J. Burns & Co. Mr. Arnold for 11 years had been captain of a fishing schooner, going out after mackerel and all kinds of groundfish. They are one of the 27 firms that first leased T Wharf and are the only firm in its original shape that is still in the business. In 1901 they admitted to the firm Arthur L. Parker, who had been their bookkeeper for many years, and all three are natives of Duxbury, Mass. When the Gloucester Fresh Fish Co. started in business and withdrew its vessels from Boston, Arnold & Winsor were the most earnest in their opposition to such a scheme and at once started the building of a fleet of fishing vessels to take the place of the Gloucester vessels and so insure a supply of fish for Boston markets. They at one time owned interests in and controlled the movements of 14 of the finest fishing schooners ever hailing from Boston. In 1911 they incorporated, with Mr. Arnold as president, Mr. Winsor as vice-president, and Mr. Parker as treasurer. Mr. Arnold has been president of the New England Fish Co. and a director in the Canadian Fish & Cold Storage and the Northwestern Fisheries since 1908, and a director for 27 years in the New England Fish Co. Mr. Parker is a director in the Commonwealth Cold Storage, and Mr. Winsor a trustee of the New England Fish Exchange, and was also a trustee of the T Wharf Corporation.

The Atlantic Halibut Co. The halibut producing industry was first organized in Gloucester, Mass., about 1873 or 1874, and has been in existence over forty years. It was first known as the Union Halibut Co., and afterwards as Stockbridge & Co., and was thus continued till 1879, when a syndicate was formed comprising the owners of fishing vessels for mutual protection and to increase the sale of such fish. They took over the business of Stockbridge & Co., and each firm was obligated to have two of its vessels engage in the halibut fishery the year round, and also agreed among themselves to market first the product of their own vessels, afterwards purchasing the fish from wherever they could be bought the most reasonably. This company was composed of eleven firms, embracing the best and strongest firms in all Gloucester. They did for years a very large and successful business, selling their fish in Boston, New York, Philadelphia and other large cities, confining themselves strictly to the wholesale trade. In 1895 the company became incorporated and many of the firms who owned the

vessels withdrew from the business. Since its incorporation it has always maintained a branch office in Boston. In 1909, owing to the great change in the production of the Atlantic Ocean halibut and the quantities of that kind of fish which were being shipped from the West Coast, owing to the depleted condition of the Atlantic side, it was deemed advisable to separate the company into two parts, and the Atlantic Halibut Co., of Boston, was incorporated with Hon. David I. Robinson, president, Harry L. Belden, treasurer, and Gardiner Poole, manager. They were located at 126 Atlantic avenue, and there remained until removing to South Boston. When shipments from the West Coast began to come they were among the pioneers, as well as in the shipment of fresh salmon to Boston and New York. They have a fine location in Seattle and are connected with the most modern freezers and packing establishments on the West Coast. They purchase their fish direct from the producers. They are a close corporation and their stock is very securely held in few hands. They are located at No. 40 on the new pier. They move along side by side with their largest competitor, the New England Fish Co., without the slightest friction, exhibiting a rare condition of friendliness in business relations.

John Burns Co. Successor to J. Burns & Co. and John Burns, Jr. Mr. Burns, Sr., commenced operations on old Commercial Wharf under the name of J. Burns & Co. in 1872, being the sole owner. Previous to this Mr. Burns had a wide experience on the sea from a very early age, which carried him to many foreign lands. When most of the wholesale dealers moved to T Wharf Mr. Burns went with them, obtaining store No. 1 at the head of the pier and remained there until 1910, when he retired from business. At this time the business of John Burns, Jr., and J. Burns & Co. were incorporated under the name of John Burns Co., a Massachusetts corporation, John Burns, Jr., president, William F. McKeon treasurer. This firm, as well as all other firms in the wholesale business, removed its business to the new fish pier at South Boston April 1, 1914.

Baker, Boies & Watson They started as a partnership in 1901, consisting of B. B. Baker, A. S. Boies and John and Albert Watson, at 25 T Wharf, in a general wholesale fish business. The two Watsons had formerly been masters of fishing schooners, and both had consider-

125

able experience in yachting circles, John having been sailing master of a sailing yacht and Albert had at one time been mate of the famous *Constellation*. Mr. Baker was the son of the man, and connected with him, who for so many years had charge of the scales and the weighing of fish at T Wharf. Mr. Boies had been a general salesman. They continued as a partnership until 1908, when they incorporated, with Albert Watson, president; John Watson, treasurer, and George Grueby, who had previously been with L. B. Goodspeed & Co., but who had just joined this company, as clerk. They moved to the new pier with all the others last April, where they are now located. They are doing a safe and sound business, catering more to local trade and near-by cities than to places farther away. They handle fine goods and seek for good trade. They are earnest workers and deserve all the success they are likely to attain.

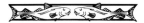

Boston Fish Company, Inc. This company was organized in 1874, started business on Commercial Wharf, located at Nos. 81 and 82. It was composed of J. Frank Bunting, who was a son of the Mr. Bunting of the old Bunting & Emery firm, and George Watts. They continued together for about two years, Mr. Bunting being a conservative, cautious man, while Mr. Watts, given somewhat to speculative plunging at times, left the firm and went with the Gloucester Fish Co., which firm bore no relation to the present Gloucester Fresh Fish Co. Mr. Bunting carried on the business alone from that time on, and removed to T Wharf in 1884, still continuing business alone until his death, which occurred in March, 1913. Mr. Watts died some few years ago. During the last few years of his life, Mr. Bunting suffered greatly from rheumatism and became almost a pathetic figure as he came and went, putting up a brave struggle for life as against a dreadful and painful disease. The house dealt largely, as it does to-day, with New York and Philadelphia, and other far-away points, and at the same time, has a goodly local trade. After Mr. Bunting's death the firm incorporated, with Mrs. Bunting president and treasurer; Robert E. Rorks, vice-president, and Florence M. Bunting as clerk. Mr. Rorks has been connected with this house from its organization.

Bunting & Emery Co. The above firm is one of the oldest in the history of the fish business now remaining under the same name. It was organized in 1867 and located on Commercial Wharf, Samuel M. Bunting and Freeman Emery being the original partners. Mr. Bunting formerly had been in the firm of Terrell, Bunting & Co. He came from Swampscott and had been a fisherman, and had first been connected with A. Pierce & Co. at Commercial street. Mr. Emery had been with the old reputable firm of Brown, Seavy & Co., which had done a large business, both in fresh and salt fish. These two men brought to their new firm quite a large and varied experience. They did a large and profitable business at that old location for years. Mr. Bunting died in 1882, and then Marshall F. Blanchard, who had entered the firm as bookkeeper several years before, was admitted to a partnership. In 1884, in company with all the other dealers, they came to T Wharf and located at No. 8, and there remained till this last April, when they removed to the new fish pier, and are now located at No. 7. They are still doing a good and profitable business. Mr. Emery died in 1897, and his son, Frank W. Emery, took up his father's interest in the business and remains in the firm still. Mr. Blanchard was elected president of the T Wharf Association in 1900, and remained in that position till the present time. He is a director in the New England Fish Co. and also in the Boston Fish Market Corporation. The firm incorporated in 1912, when Benjamin R. Atwood, who had been bookkeeper with the firm for 32 years, became a stockholder in the company. Mr. Blanchard is the president and Mr. Emery the treasurer.

The Bay Fish Co. Seventeen years ago there came across the sea a young man seeking his fortune in this new land. He drifted around the wharves and found employment with the firm of E. A. Rich & Co. This young fellow was a willing worker; he quickly made himself useful to Benjamin Rich, the buyer for the firm, and before many realized it, Jack O'Hara was buying the goods himself for the firm. Some five or six years later another brother, Patrick O'Hara, appeared and in two years more William O'Hara came over and joined the other two. Each was a worker; each of them saved some money, and deciding that they could do better for themselves, in 1908 they incorporated themselves as The Bay Fish Co., with John F. as president, and the others, one clerk and one treasurer. They located at 116 Atlantic avenue in a little store, 16 by

40 feet. This Bay Fish Co. was the parent of all these other stores. In 1910 John F. saw the opportunity of the Witherell Fish Co. and purchased it, going there to take charge of it, and leaving William in charge of the Bay Fish Co. In 1912 the Long Wharf Fish Co. was acquired, and Patrick O'Hara was placed in charge of that store. In 1913 the old Coleman home was for sale, and that was acquired, and not having brothers enough to go around William J. Dodd, son of the man who worked so long there, has been placed in charge of the Coleman house for the present. John F. O'Hara is president of all these stores in which he is mentioned excepting the Hamilton-Prior Fish Co. The Bay Fish Co. occupies No. 3 on the new fish pier, and under William O'Hara's guidance and supervision is doing a good business, and we see no reason but what he will continue to do so.

The Booth Fisheries Co. This is the outcome of the A. Booth & Co., which was purchased from William J. Emerson, who had been one of the largest and most thriving fish dealers in Boston for many years, and who sold out to A. Booth & Co., in 1898, and remained acting its manager for this new firm until 1901, when he retired from active business. M. P. Shaw, who had been with both Emerson and the A. Booth & Co., assumed the management and ran it very successfully for ten years. The firm put improved methods into the business very early after purchasing the same. They installed the first ice-crusher run by power, and built in their store the first private cold storage rooms for the handling of their own stock. They did a large and remunerative business. In 1911, Mr. Shaw purchased the control of the E. A. Rich Co., and resigned his position with the Booth Co., when this business was entirely reorganized and the name was changed to the Booth Fisheries Co. After Mr. Shaw's retirement, John Gourville was installed as manager and continued with them until the beginning of this present year, when he resigned and went back to his former position, and Samuel McIntype, who for years had been with the John R. Neal Co., now has the management of the firm. They occupy the largest area of any firm on the New Pier and again have their own cold storage plant installed. The business in Boston is but one of the branches of a large chain whose headquarters are in Chicago. They have large plants on the West Coast as well as on the Great Lakes, they operate steamers and tugs in the halibut fisheries, and have five large cold storage plants in various parts of the country. They

have carried on for years a smoke house and a salt fish curing place in East Boston, and have lately organized a salt fish plant in Gloucester under the management of J. M. Kincade, who has been in their employ for thirteen years. J. C. Wheeler is the district manager for the Booth Fisheries interests in the East and maintains his headquarters at their office on the New Fish Pier, from which he travels as occasion requires. They are the largest fish firm in the United States, if not in the world, under one general management.

A. G. Baker Born in Fall River, Mass., in 1870, at the early age of ten, Mr. Baker began peddling fish with a basket between school terms and before and after school hours, and this he continued until he was fourteen. In 1884 he went to work for a dealer in his native city and continued with two of them until he was eighteen, and then in 1888 started for himself in his own town. It is quite amusing to hear him tell how when so young he went on to New York, and there made the acquaintance of the firm, now Chesboro Bros., located at No. 1 Fulton Market, with whom he has traded ever since, owing to the kindly manner in which they met his first advances. He remained in Fall River until he was twenty-four years old, and then in 1894 he decided that Brockton, Mass., offered goodly promises for a wholesale fish house. He accordingly located there on Freight street, close by the railroad station, where he at once branched out into quite a business. He went regularly to Boston, buying his goods both there and in New York, and got them to his market as quickly as possible, where he carried on quite a trade with markets smaller than his own who did not care to buy their goods in large lots. In 1902 he found Brockton too small a field and consequently came to Boston, located at 102 Atlantic avenue, where he entered into the full-fledged wholesale business. He remained there until April 1, 1914, continually increasing his business, when he removed with the other dealers to the New Fish Pier. To Mr. Baker belongs the honor of buying the first store on the largest and most sanitary fish pier in the world, and is located at No. 1 on the new pier. He prides himself upon handling the finest goods he can buy, and tells you laughingly that his is the house where poor quality is unknown.

Coleman, Son & Company This is one of the very oldest firms in the fish business. In 1864, on Commercial Wharf, William Coleman and P. J. Coleman, his son, started the wholesale fish business, which was far different from the manner in which the wholesale fish business is conducted to-day. A few years later, Thomas R. Stinson entered the firm and took charge of the office and the finances. They continued on Commercial Wharf until 1884, when, with the other dealers, they moved to 27 T Wharf, being the farthest down on the pier of any dealer. They conducted a business not so large and varied as many other houses, but a business that, in the earlier days, paid them remarkably well. In 1872, John Dodd, entered their employ and has continued with them until the present time, making a continued service of 42 years, one of the longest of any employe in the fish business. On Dec. 15, 1913, this old firm of so many years' standing, owing to the death of all the originals, sold out to John F. O'Hara, president of the Star Fish Co., who controls several other stores. The firm then became incorporated under the same name, becoming Coleman & Son Co., of which Mr. O'Hara is president and treasurer, and William J. Dodd, manager in charge. They are now on the new fish pier, No. 38. Under Mr. O'Hara's energetic management the firm is picking up new life and new trade, and we expect to see it very shortly take its place in line with the other stores that Mr. O'Hara controls.

Freeman & Cobb Co. Freeman & Cobb Co. were incorporated in 1902, with William I. Atwood, president, and W. Elmer Atwood, vice-president; N. D. Freeman, treasurer, and Irving M. Atwood, secretary and manager. This house is really the consolidation of three firms, two former wholesalers and one commission dealer. The original firm of Freeman & Cobb was composed of George Freeman and Solomon T. Cobb. They had both been employed on the old Commercial Wharf, but the make-up of this company is so large we cannot trace them back through all their early history. Freeman & Cobb did a good business and had a neat, snug trade. Mr. Cobb died recently. N. D. Freeman & Co. begun like so many others on Commercial Wharf. They were located at No. 24 T Wharf after their removal there. Atwood & Co. were commission dealers, which business was established by Capt. John Atwood, of Provincetown, father and grandfather of the Atwoods above mentioned. When the National Fish Co. was formed on T Wharf, Freeman & Cobb, and

N. D. Freeman & Co. both entered its fold, and remained with it until its dissolution; then finding that strength lay in union and co-operation, they decided to merge their two businesses and assumed the name of Freeman & Cobb; the Mr. Freeman of the old Freeman & Cobb having died. They took a store, No. 3 T Wharf, both firms moving into that store, when they associated with them Atwood & Co., thus uniting three very able firms, and three exceedingly capable men. While Mr. Freeman is treasurer, he takes no active part in the management of the business. W. I. Atwood is treasurer and manager of the Consolidated Weir Co., the Cold Storage Co. of Provincetown, and also of the Atwood & Payne Co. of Gloucester. Irving M. Atwood is a director in the Commonwealth Ice and Cold Storage Co., and a director in the Consolidated Weir Co. N. D. Freeman is president of the Bay State Fishing Co., vice-president and trustee of the Boston Fish Market Corporation, director in the New England Fish Co., and of the old T Wharf Corporation as well. This house does a large and successful business, with the lines of trade which each house has turned into the general business and with the supply that Atwood & Co. continually have to draw from, they are at all times amply supplied with stock of all kinds ready for sale.

Warren Fitch Co. This company was organized June 1, 1904, with Mr. Fitch president, at 100 Atlantic avenue. Mr. Fitch had been employed for many years with the old firm of Freeman & Cobb on T Wharf, and when the old firm of Freeman & Cobb became merged with the house of N. D. Freeman & Co., Mr. Fitch also followed into their employ, and was a very capable and interested foreman for them for several years. Born in Boothbay, Me., coming to Gloucester in early life, he had pursued all the various branches of fishing, both hand lining, trawling and seining for mackerel. Getting somewhat restive while with the Freeman & Cobb Co., he decided to start out for himself, and organized this company, as above stated. From the very start he has done a good business. He is very careful in his buying, handles only the best stock he can find, and caters to the trade who use it and are willing to pay for such goods. He is energetic, a good worker, and although he was at the extreme northern end of the fish trade on Atlantic avenue, he managed to pick up and retain good customers. He is a pleasant man to meet, at times very entertaining, but if one sells him poor goods, and does it purposely, his trade with Warren Fitch is at an end. Mr. Fitch is right in the prime of life, and there is no doubt that he will attain all the success which hard work and careful oversight of his business can give.

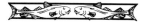

Fulham & Herbert This is the firm who really succeeded to S. H. Jackson & Co. Stephen H. Jackson was employed for years by Bunting & Emery, on Commercial Wharf, and at the removal of Bunting & Emery in 1884 to T Wharf, left their employ and started business for himself on Atlantic avenue. Both Mr. Fulham and Mr. Herbert had been employed by Mr. Jackson for many years, and at his retirement purchased the business, started in for themselves, and despite all the incorporations that have gone on about them, remain a partnership still. They make a grand team together and are first-class workers, each taking his allotted place, Mr. Fulham attending to the buying from the fishing vessels that come in to the pier, and Mr. Herbert taking charge of the floor of their wholesale house, and supervising the putting up and shipping of the various orders. By so doing, they have succeeded amazingly well. They can be classed as general wholesalers and jobbers. Their trade is largely barrel orders, but they carry the stock, and are able and willing at all times to furnish box goods. When the projected removal to South Boston was in the air, they became earnest supporters of the same. At the last election Mr. Fulham was made president of the New England Fish Exchange. This exchange has done a very good work, both for the dealers and the fishermen, securing a fairer market and a better means of purchasing fish, and Mr. Fulham will, without doubt, keep it in the front rank it has so long enjoyed. Perhaps few firms are better deserving of success than N. L. Fulham and John J. Herbert.

Gloucester Fresh Fish Company For many years, especially during the winter, Gloucester vessels had been bringing their fish into Boston for market. They purchased their stores, bait, ice and the men spent their money largely in Boston. During the summer of 1897 the Gloucester merchants and vessel owners concluded this was all wrong, and they set about devising some plan which should bring their vessels home with their fish and thus give Gloucester owners and storekeepers the benefit of all this trade that was left in Boston, so they organized the above named

company, with Hon. William H. Jordan as president, Capt. William Thompson as vice-president, John J. Pew treasurer, the Hon. David I. Robinson general manager. The merchants of the city subscribed and paid in a cash capital of $75,000, and the owners of the vessels furnished them for the production of fish, taking stock for the same. They started operations in November of the same year. They then had 214 vessels and boats fishing for and controlled by them. The weather was fine and fish were plenty. Boston was their main market, and C. K. Sullivan was their manager in Boston, and there were shipped to him one day 1,000 boxes of fish, containing a total of half a million pounds, and 100 boxes were sold to one firm and 150 to another. After a while they opened two stores in Boston, which later merged into one on Atlantic avenue. In 1899 Mr. Robinson ceased as manager and Edward J. Livingston succeeded him and remains as manager at the present time. In 1899 Howard Steele became treasurer. The company has had four presidents, William S. Jordan, Thomas J. Carroll, C. W. Luce and W. C. Brown, who is now in office; David B. Smith, now vice-president, with Mr. Steele and Mr. Livingston, all re-elected last October.

L. B. Goodspeed & Co. In 1869 Mr. Frank C. Goodspeed entered the employ of that old fish firm, John Marston & Co., located at 207 Commercial street, Boston, and in 1873 his brother, Edwin S., also entered the same employ. The firm of John Marston & Co. consisted at that time of Mr. Marston and Augustus Winsor. In about 1879 Mr. Marston left the firm, but it still continued under the old name, and was one of the leading houses in the business, for when they came to T Wharf and located at No. 16, in 1884, Mr. Winsor became the first president of the Corporation. They did a good business and it is stated by some of the older dealers that Mr. Winsor's permission to do certain things had to be carefully obtained. In 1891, the firm dissolved, and it became known then as F. C. Goodspeed & Co., the two brothers mentioned above. They continued on in this way until 1894, when a younger brother, L. B. Goodspeed, who had been employed as bookkeeper by the old firm of Arnold & Winsor, left that position and joined hands with his brothers, when the firm name was changed to L. B. Goodspeed & Co., and remains so to the present time. They do a good business, branching into all the various lines of trade, the local barrel trade and the box trade of the larger cities. They have extensive dealings with

New York, Philadelphia and Baltimore, and are extremely genial and pleasant people to deal with. L. B. Goodspeed was elected treasurer of T Wharf Association in 1891, a position he continues to hold. He is director in the Commonwealth Ice & Cold Storage Co., the Boston Fish Market Corporation, the New England Fish Co., president of the Northwestern Fisheries Co., vice-president of the Canadian Fisheries Co., and a director of the Bay State Fishing Co. This firm deserves all manner of success.

F. E. Harding & Co. This old firm was established by John Harding, on Commercial Wharf, as long ago as 1860, and is one of the real pioneers in the wholesale fresh fish business. Mr. Harding was quite an elderly man, and later on Frank E. Harding, his grandson, whose name the firm bears to-day, took over the business from his grandfather and conducted it alone till 1877, when he associated with George H. Clark, and in 1884, they moved to T Wharf and located at store No. 30. F. E. Harding died in 1886, leaving his half interest in the business to his heirs. In 1888 Elmer T. Randlett came fresh from a commercial school and took the position of bookkeeper with Mr. Clark, and remained in that position until 1894, when he, with Samuel E. Rice, who had been a buyer for the firm, were admitted as partners with Mr. Clark. Mr. Clark died in 1902, and the two other partners bought Mrs. Clark's interest, when they continued it along under the same name. In 1908, they incorporated and in 1912, Mr. Rice died, when Mr. Randlett purchased the stock and has since continued the business alone. He moved to the New Pier last year and it has gradually increased from a local trade and barrel orders into a heavier and larger trade with New York and Philadelphia. Some two years ago the highest point the business had ever reached was attained. Mr. Randlett is a director in the New England Fish Exchange, in the New England Fish Co., and in the Northwestern Salmon Co. Pursuing the same course in their business as heretofore, directing with the same force and exerting the same energy, it is small wonder that their business moves up in an ever-increasing line.

Cassius Hunt Company Incorporated in January, 1912, with Cassius Hunt, president; George Hunt, vice-president, and J. Henry Hunt, treasurer. The house is really a successor to the old C. Hunt & Co., which was organized by Cassius Hunt and Edwin Hunt, brothers, twenty-nine years ago, and located on South Market street near to Atlantic avenue. Both the brothers had previously been employed by William Prior, Jr., in Faneuil Hall Market in the retail stand now occupied and run by Prior & Townsend Co. Two years later, Charles Payson was admitted as a partner, and took charge of the office and the finances. In 1906 Edwin Hunt sold out his interest to George and Henry Hunt. Edwin Hunt died in 1912, six years after retiring from active business. In January, 1912, Mr. Payson retired from the firm, selling his interest to George and Henry Hunt, who by this means became largely interested in the ownership of the business. Cassius Hunt died in June, 1914, and the business was then incorporated. The present firm of C. Hunt Co. removed from South Market street to the new fish pier on April 1 of the present year, and are located at No. 17 on said pier.

Haskins Fish Co. Moses and Leander Haskins were brothers, coming from Rockport to Boston and were partners together in the wholesale fish business previous to the dealers coming to T Wharf, over thirty years ago. When the fish dealers decided to lease the T Wharf property, Leander Haskins was one of the parties who signed the lease and they started in business on T Wharf at No. 18. They became interested early in their business life in the manufacture of glue and isinglas, and as the years passed, gave more and more of their attention to those industries. In 1892, they sold out their business on T Wharf, to Charles A. Grant and Gamaliel Rich, both of Winthrop, Mass. Mr. Rich, coming from Provincetown, had always been identified with the fisheries, and had been employed by the Haskins brothers for several years. They did a large business in what is termed a "box" trade, shipping cod, hake, pollock, etc. to New York, Philadelphia, Baltimore and to Western cities. They did not favor the project of removing to South Boston, and some four years ago, sold out their business to John N. Fulham, who had formerly been with his brother, of the firm of Fulham & Herbert, and is now located at 26 Fish Pier. Mr. Fulham is an enterprising young man of good habits, and while carrying on business in much the same line as the old firm is giving more of his attention to the barrel trade or filling small orders for retail dealers. He still retains the old name, Haskins Fish Co.

Henry & Close Joseph Maddock came early in life from England, and for many years was bookkeeper for the old firm of Taylor & Mayo, near the end of T Wharf. After many years of service, he, in company with Leonard A. Treat, salt fish dealer, formed the J. Maddock Co., located on Atlantic avenue, and after a time moved to Long Wharf, where they carried on a general wholesale fish business. Mr. Maddock was a bright, enterprising young man, and at times, somewhat of a plunger. After a while, he engaged Samuel S. Close, who had been the foreman for Ferguson teaming business, as foreman for the Maddock Co., and when Mr. Maddock in 1903 went West, and gave up the Boston business, Mr. Close became manager and later, together with Alfred F. Henry, who had been buyer for F. H. Johnson, purchased the business and called it as above, Henry & Close. They make an excellent team. Mr. Henry attends to the buying from the fishing schooners, and takes the outside end, while Mr. Close makes a remarkable housekeeper and tends the store with all diligence and great care, seeing that everything goes out in proper order. In 1907, they moved to Commerce street, and there remained till April last, when they moved and are now located at No. 20 on the new fish pier.

Wm. Haskell Co. This firm, one of the very oldest in the business, was established in 1859, under the name of Parsons & Co., consisting of Wm. Haskell, Mr. Langford and Mr. Parsons, and was located on Commercial Wharf. The year 1859 is a long while ago, the fresh fish business in Boston was largely in its infancy and had made no such developments as existed years later. In 1874 Mr. Haskell purchased the interests of his other partners and it then became known as Wm. Haskell & Co. In 1884, when the dealers came to T Wharf, this firm located on the property belonging to T Wharf, known as Commercial street, just above Atlantic avenue, where with his two sons as helpers, he conducted a safe and frugal business until 1898, when he died at the advanced age of eighty-nine years. After that Henry Haskell, the younger son, took over the business and conducted it in the same quiet, careful way as his father had done until March, 1914, the date of his own death. The business now belongs to his estate, and was incorporated in January, 1915, with Kenneth A. Haskell, president and Joseph O'Connor, treasurer. This house, although never doing a large business, certainly enjoys the distinction of great age, for no firm has been found that can be called strictly a

fresh fish house that started business so early as 1859. There was then no Atlantic avenue thought of, and the Civil War had not yet thrown its shadow across the country. The elder Mr. Haskell was a pleasant, genial old gentleman, and his loss was greatly deplored among the dealers.

George M. Ingalls & Co. The firm of Prior, Ingalls & Co. was organized Nov. 24, 1883, with location at Commercial Wharf. The firm consisted of George P. Prior and George M. Ingalls, both of whom had been in the employ of P. H. & William Prior, Jr., since 1872, where Mr. Prior was the buyer and Mr. Ingalls the bookkeeper. After organizing their partnership they remained on Commercial Wharf, doing a general wholesale and commission business until 1884. Then, when the dealers went to T Wharf, they removed to 124 Commerce street, which at that time and now, is part of the old T Wharf property, and there the business went on in the usual routine until April last, when they, in company with all the other dealers, went to the new fish pier, where they are now located at No. 5. The firm has always done a good business and has prided itself more on handling good stock than in the large amount of goods it could handle. George P. Prior died Aug. 30, 1908, when Mr. Ingalls took over all his interests and now conducts the business alone. He has always aimed for good help, and some of his associates rank very high in the lines of work they are called to do. Mr. Ingalls was a trustee of the T Wharf Association, is vice-president of the Boston Fish Market Corporation, is a trustee and treasurer of the Commonwealth Ice & Cold Storage Co., is a member of the finance committee of the New England Fish Co.

Lombard & Curtis Unlike the many firms which have incorporated their business during the last few years, this house still retains the old form of partnership. Originally it was composed of Oliver C. Lombard and George Curtis, and continued so until the death of Mr. Curtis some years ago, when his son, Edgar F. Curtis, succeeded to his father's interest and the business continued under the same name as before. This firm started in 1884, when the dealers removed from Commercial Wharf to T Wharf, and were located at No. 17, and there remained through 30 years, until their removal in April last to the New Fish Pier. Earlier in life Mr. Lombard and the elder Mr. Curtis had both been employed by the same firm, Snow & Fuller, one of the oldest houses in the business. They are now located at No. 45 on the New Pier. They are more of the old-school type than many of the more modern and younger firms, attending closely to their business, doing much of the work themselves, careful in their buying, and still more careful in their selling. They are a very conservative and prosperous house, pleasant people to meet and pleasant to deal with, and deserving of all the success that they had attained and are still likely to attain.

John R. Neal Co. This company, which was started by John R. Neal, is one of the oldest fish concerns in Boston. It was first opened for business in 1878 and located at 28 Commercial Wharf. When T Wharf was opened as a permanent fish market the firm moved to store No. 22 T Wharf, and later expanded and occupied stores Nos. 21, 22 and 23.

In 1888 Frank W. Neal became associated with his brother, under the firm name of John R. Neal & Co., and during the same year operated the first smoke houses in Boston, making finnan haddies, afterward buying their own property in East Boston and erecting model smoke houses for this department of their business. The Neal brand finnan haddies took the highest award in 1893 at the Chicago World's Fair.

The members of this firm were pioneers in investigating the halibut industry on the west coast, which industry has since grown to large proportions. John R. Neal was first president of the New England Fish Exchange, and was active in the development of the new Fish Pier. Frank W. Neal still remains as manager of the company, while John R. Neal has retired from active business.

John R. Neal Co. was incorporated under the laws of Massachusetts in 1909, and now occupies store 33, Boston Fish Pier, the largest store on the pier, with their own freezing and refrigeration rooms.

R. O'Brien & Co. The old reliable firm of R. O'Brien & Co., was established by Robert O'Brien in 1865. At the age of fourteen he entered the employ of Brown, Seavey Co. He continued with them and afterwards formed a partnership with Wm. J. Emerson. The business was conducted on Commercial Wharf, the firm being known as Emerson & O'Brien. This continued for five years, when Mr. O'Brien severed his connection with the firm and purchased the interest of John S. Wright. Later on he went to Rockport, Mass., to purchase fish and in 1898 he established the firm of R. O'Brien & Co., associating with him, his sons, Wm. J. and Daniel J. At the death of Robert O'Brien, his sons succeeded him. They have operated the business successfully since. When the New Fish Pier was first projected and the New England Fish Exchange formed, Wm. J. O'Brien was elected vice-president. He is now president of the Boston Fish Market Corporation, and also the Commonwealth Ice & Cold Storage Co., both of which have profited by his judgment and executive ability in placing these properties on a satisfactory basis.

F. J. O'Hara & Co. The above company is officered by F. J. O'Hara, Jr., president, and F. J. O'Hara, Sr., as treasurer. Mr. O'Hara, Sr., started life at thirteen years of age as a cabin boy on the ship *Merrimac*, plying between Liverpool and New York, making several trips across the ocean. He gradually worked into fishing out of Boston. He was in the Navy in 1865, at the close of the Civil War, and was at New Orleans at that period. That seemed too quiet for him after the tumult of war and he shipped on a Norwegian vessel to go to Genoa for $150 in gold. He was at one time one of the crew on a ship which took the first load of ice shipped from Boston to Cape Town, South Africa. Tiring of the sea, and deciding to go into the fish business, thirty-seven years ago last April, he started in at the store, 112 Atlantic avenue, having as his first partner Mr. Charles Leonard, now deceased, who remained with him about one and a half years. Mr. Leonard retiring, Richard Manchester became his next partner for a year or so, when he wished to buy out Mr. O'Hara, but the result of which was just the contrary. Then the business was moved to 118 Atlantic avenue, where he remained until last spring, when, with the other dealers, he came to the Fish Pier, and is now located at No. 13. They have always done a large and very successful business, always ready to oblige, equally ready to sell. Twenty-five years ago, Mr. O'Hara moved to Winchester, where he has an elegant estate bordering on the State Boulevard. He is a director in the Boston Fish Market Corporation, in the Cold Storage, and a trustee in the New England Fish Exchange. He has served as selectman in Winchester, is a director in the Co-operative Bank in Winchester and a director of the Columbia Trust Co., in Boston.

The Ocean Fish Co. In 1908 the John R. Neal Co., then at the height of their business on T Wharf, desiring a larger share of fish from the boats than it was always easy to get, conceived the idea of taking over store No. 20, which adjoined their own, and calling it the above name in order that it should apparently stand by itself, but which at the same time belonged to and was controlled entirely by themselves. During all that time and before Mr. Frank R. Neal was buyer for the house and also purchased goods for this latter acquisition. The Neals soon set this new store going at the same rapid gait they had traveled in the old store, and a goodly lot of fish went in and out of the Ocean Fish Co. whenever there was anything doing on T Wharf. When, in 1912, the Neals sold out their business to the Booth Fisheries they sold the Ocean Fish Co. to Frank R. Neal, who continued it until his removal to the New Fish Pier last April, where he is now located, very near the end of the pier on the right-hand side going down. This Mr. Neal is a smart, active young man. He has the courage of his convictions and dares to act at all times. He has an established fish place in Provincetown, Mass., which he started in 1908, where he is ready at all times to buy fish of all descriptions, either from the local fishermen or the fishing vessels which go in there for harbor and ship them to Boston, New York or Philadelphia, as the market may promise best. He carries on a general wholesale business and has an extensive commission business, receiving shipments from all directions. He takes strong chances, is perfectly willing to work and is deserving of all the success he may attain.

P. H. Prior Company This company is really a successor to the old house of P. H. Prior, well-known to all the dealers in Boston and generally throughout the country. Mr. Prior originally started business in Quincy, peddling fish very early in life. Becoming dissatisfied with the small business and looking for a larger field, he came to Boston, entered into partnership with his cousin, William Prior, Jr., in Faneuil Hall Market. About 1872 he started in the wholesale business wholly by himself at 30 Commercial Wharf, and remained on Commercial Wharf until 1884, when he, with the other wholesalers, came to T Wharf, he being one of the signers of the lease which rented the wharf as a fish property. For many years he was president of the Union Ice Co., being one of its incorporators, and also was president of the T Wharf Herring Co., both

of which positions he held until the time of his death, which occurred in April, 1900. He was a man whose word was as good as his bond. After his death the firm continued as a partnership until 1910, when it was then incorporated. George A. Wyer, of Portland, who had previously acquired an interest, becoming president, and David F. Choate, previously the bookkeeper, becoming treasurer, and Elmer E. Prior, manager. They are now located at No. 29 on the new fish pier. Mr. Choate has recently been elected a director of the Fish Exchange.

B. F. Phillips Co. Herbert F. Phillips is president and Frederick G. Phillips is treasurer of the company above named. George W. and Benjamin F. Phillips were two stirring and energetic brothers, living in the town of Swampscott, Mass., before it had attained such celebrity as a beach and summer resort as it to-day enjoys, but when fishing was carried on there as one of the means of livelihood. In those days the boats, and some of them quite large boats, went out early in the morning, returning in the afternoon with the fish, which were carefully selected and prepared for the Boston market, to which they were taken over the road in teams, and these two brothers were very busily engaged in this line of business. From their frequent trips to Boston they concluded that they could do far better to establish a wholesale fish store in the city and have the goods shipped to them; consequently they located on Commercial street, in the rear of an old blacksmith's shop, where they at once succeeded in doing a very good business. They early earned the reputation of handling good stock and the Phillips Beach fish soon gained a commendable notoriety. Later, in 1884, they moved to 20 T Wharf, their business growing apace. In 1892 George W. Phillips died, and the firm name became changed to B. F. Phillips & Co. Meantime they had held and increased their reputation as large handlers of codfish, especially steak cod, which went largely to New York City. George Phillips, Jr., who had worked with them, died in 1894, and Benjamin F. Phillips died December 23, 1896. The two named gentlemen at the head of this article, respectively, the sons of Benjamin F. and George W., assumed control of the business, and have carried it on very successfully. Both are strong, energetic business men, with abundance of courage always approachable, willing to impart any information at their command and are doing a grand, good business.

E. A. Rich & Co. In 1864, on Commercial Wharf, E. A. Richard and A. F. Rich, brothers, started in business as A. F. Rich & Co. Something over a year later they bought out part of the Holbrook & Smith Co., when George Andrews and Joseph Rich came into the firm, which they then called Andrews, Rich & Co. They ran on in this manner until 1872, when E. A. Rich took sick, drew out from the firm and went into the woodworking business. Tiring of the wood manufacturing business, and knowing the fish business so well, in 1878, together with another brother, Benjamin F. Rich, he started again a fish firm on Atlantic avenue, and remained until 1884, when the firm came to T Wharf, where they remained until the removal to the New Fish Pier. About three years ago E. A. Rich sold out his interest in the firm to Maurice P. Shaw, who had formerly been with William J. Emerson for 18 years, and who followed with the Booth Co. after Mr. Emerson sold the business. Then the E. A. Rich Co. became incorporated, with Mr. Shaw as secretary and treasurer and Benjamin F. Rich as president. The firm has always been an adventurous and progressive one, buying only the best and finest goods and catering to a trade that was glad to get such goods and willing to pay for them. Mr. Rich was an exceptional buyer. Mr. Shaw is a worker and a hustler. He is treasurer of the Boston Fish Market Corporation, secretary of the New England Fish Exchange, a director of the New England Fish Co. and was a trustee of the old T Wharf Corporation.

A. F. Rich & Co. In 1866 A. F. and E. A. Rich, two brothers, formed a partnership to carry on the wholesale fish business and located on Commercial Wharf. In the year following they bought out Smith & Co., of the firm of Holbrook, Smith & Eldridge, and removed to stores, 9 and 11 on the same wharf, at which time they were known as Andrews, Rich & Co. In 1880 A. F. Rich bought out the interests of all the others, E. A. Rich having previously retired from the firm on account of ill health, and the firm again became A. F. Rich & Co., and has remained so until the present day. Mr. Rich early became interested in the proposed removal to T Wharf, and as he is a very positive man became quite a factor in such movement, and in 1884, in company with all the other wholesale dealers removed to No. 13, T Wharf, where he remained during the entire thirty years of the lease of that property. In 1872, the New England Fish Co., at that time known as the New England Halibut Co., having been pre-

viously organized with its main office in Gloucester, Mr. Rich was elected its secretary, and also acted as assistant treasurer for the Boston end. In 1902, when the company was incorporated under the name it bears to-day, it having previously been an association, Mr. Rich was elected its permanent treasurer, a position he holds to-day. At that time he left the old store at No. 13, took up his office quarters with the New England Fish Co., at 24 T Wharf. This store is now run by Ernest F. Rich, his son, who moved with the dealers to the new fish pier and is now located at No. 2. They do a general wholesale business. Ernest Rich is a pleasing young man, quiet and unassuming, but attending strictly to his business and from conditions we judge him to be more successful each year.

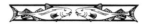

H. A. Rich Company H. A. Rich Co. are really the successors to the firm of Rich Bros., who started business at 37 and 38 Commercial Wharf about 1886. The firm was then composed of Joseph S. Rich. John L. Rich, brothers. Joseph Rich had been employed for many years by Bunting & Emery, one of the pioneers of the wholesale fish business in Boston. Feeling sure his opportunity was at hand, Joseph Rich induced his brother John, who had been master of a fishing schooner, to come to Boston and join him in the wholesale fish business. Associated with them was Capt. Henry Thomas, from Gloucester, who had also been a very successful fishing captain, who acted as buyer for the firm; H. A. Rich, the senior member of the present firm, was the son of Joseph Rich, and was employed as bookkeeper; John Rich, soon tired of city life and the business, and returned to Cape Cod, where he went into the fish weir business. After a while Captain Thomas retired from the business, and the firm then became Herbert A. Rich & Co., retaining the same location on Commercial Wharf, the business growing rapidly in the meantime. They remained on Commercial Wharf until the formation of the National Fish Co., organized in 1900, when becoming one of said company they removed to T Wharf. They remained with the National Fish Co. until its dissolution in 1901, when they took store No. 10, T Wharf, and remained there until April last. Some time previous to removing to the fish pier the firm incorporated under the above style, of which Herbert A. Rich is president and treasurer, Winfield S. Kendrick, vice-president, and William Nason, acting clerk in place of his father, James L. Nason, deceased.

Snow & Parker They started in business on Aug. 17, 1905, on T Wharf, purchasing the business formerly carried on by Ezra Coombs, which was the buying and selling of codfish tongues and cheeks, that is, they purchased the heads from the dealers who make up steak cod, and removing the tongues and cheeks from the head, sell them to a trade always calling for such goods. With this business they combined a general wholesale commission business in all kinds of fresh and salt fish. Mr. Snow formerly was foreman for F. E. Harding & Co., for many years, and Mr. Parker had been foreman in the teaming business for Alexander Grimes. They have made a very good hitch-up, Mr. Parker going out to seek consignment of goods, while Mr. Snow remained in Boston and attended to the selling end. They own and control three fish weirs at Chatham, Mass., from which they get many goods to put on sale. They are the only firm at the present time on the pier who own a team and do their own teaming. They are both very agreeable young men, very nice to deal with, and they believe in good goods, fair prices and polite treatment to all with whom they come in contact.

The Shore Fish Co. In September, 1905, Joseph Silva, who had worked in the retail fish business, and John Montgomery, who had been employed by the old firm of J. Adams & Co., for many years, organized themselves together as partners, started the whole fish business at 11 Long Wharf, known as the Long Wharf Fish Co. They remained at Long Wharf, doing a general wholesale business until 1908, when in that year, they moved to Commerce street. Mr. Silva had, however, previous to this, bought out Mr. Montgomery's interest, and was carrying on the business alone. He went on this way until about 1910, when the business was incorporated under the same name, but took in quite a few members as stockholders. In 1912, the whole business was purchased by John F. O'Hara, who became president and had the name changed to the Shore Fish Co. Mr. O'Hara infused into it new life, as he has in the other companies which he has acquired, and last spring came to the new fish pier, and is now located at No. 37. His brother, Patrick O'Hara, is in charge of this store, and together with George Tyner as foreman of the store, are making things more lively around their back and front door. They are building up a good, lively trade, are not afraid to buy, consequently almost always have something to sell. They will undoubtedly go on to a very successful business.

The Star Fish Co. Estey L. McKinney had a long and varied connection with the fishing industry. When first known to the trade he was in the employ of William J. Emerson, one of the largest and one of the most enterprising fish dealers in the whole fresh fish business. He was a very valuable man for Mr. Emerson, and did much to make that business what it was. In 1898 he became connected with the Gloucester Fresh Fish Co., remaining with them but a short time. He then went into the employ of P. H. Prior, and remained there until May, 1907, when he took a store on Commerce street, started business for himself, under the name of E. L. McKinney & Co. He opened that store in May, and died the following Nov. 22. The store was idle until April, 1908, when H. F. Witherell, one of the last remaining of the old Baker, Witherell & Co., who were in the fish business for years, took this vacant store under the name of the Witherell Fish Co. Mr. Witherell had previously had large experience in the fish business, built up a good trade, remained there until April, 1911, when the business was purchased by John F. O'Hara, who had afterward acquired the Long Wharf Fish Co., which adjoined this store, and at once changed the name to the Star Fish Co., and this is the headquarters of Mr. O'Hara as president of this string of stores. Mr. O'Hara is an active man, buys quick, sells quick, believes in large sales, quick sales, and if need be, quick profit. He has associated with him as floor manager, LeRoy Chase.

Story-Simmons Company This house was organized and incorporated in 1908 by Manuel Simmons, a sailmaker of Gloucester, who owned several vessels, and Jacob W. Story, of Newton, who for many years had been the valued bookkeeper with the firm of Prior, Ingalls & Co. Mr. Simmons was president, Mr. Story treasurer and general manager. They were located at 84 Commerce street. The firm continued for some time, when Mr. Simmons retired, owing to his other business, and after a while Mr. Story went back to his former employment. In 1911 the stock was purchased by interests representing the Bay State Holding Co., which had planned to institute a line of stores in other places. Previous to this, however, Arthur D. Story, of Essex, Mass., a prominent and well-known builder of fishing schooners, had been drawn into the business, and John P. Locker had become manager of the concern and remained with the company until long after the Bay State had purchased the controlling interest. On Jan. 17, 1914, the control of this stock was purchased by an entirely new gathering of young men, Simeon Atwood, Jr., becoming president, and William E. Curan, who had been manager some 10 years of the old reliable firm of Freeman & Cobb, becoming treasurer and general manager. They removed to the new pier with the others, and are located in one of the most commodious stores on the pier. Mr. Curan is a born hustler, does a good business and has associated with him Patrick Fitzpatrick. It was Mr. Curan who took charge of the exercises on the day of the final removal from T Wharf to the location in South Boston.

Taylor & Mayo The sign on the window states that this firm was established in 1866, but for the real beginning we must go further back than that, back to the time when Henry E. Taylor started in business in 1863, on Commercial Wharf. A short time after that one Henry Rogers was associated with him and they conducted the business together until 1865, when Richard L. Mayo purchased Mr. Rogers' interest and the firm of Taylor & Mayo began from that date. Mr. Rogers retiring from the business. In 1875, Joshua N. Taylor purchased the interest of Henry E. Taylor and the firm for years continued under this latter name. Capt. Richard Mayo had from boyhood been identified with the fishing industry, beginning life as a boy with the old-fashioned mackerel fishermen, in the days when they caught such fish with a hook and line. He had pursued it in all its branches, from Georges Banks to Labrador and all through the Gulf of St. Lawrence. Capt. Taylor had been a ship captain, sailing deep water voyages and had a reputation of making some excellent passages across the ocean. They made an ideal combination, Capt. Mayo being somewhat adventurous and Capt. Taylor careful and calculating. They came to T Wharf in 1884, continued together, doing a good business until the National Fish Co. appeared, when they merged with that, and at the expiration of that business they dissolved partnership, Capt. Mayo conducting the business alone until his death. After this his son-in-law, Mr. Henry Nickerson, of Winchester, took over the business for the estate, and managed it very ably and shrewdly. He brought the business to the New Fish Pier with the other dealers, where it is now located at No. 28. Since that time the business has been sold to John F. O'Hara, of Star Fish Co., but continues under the same firm name of Taylor & Mayo.

United Fish Co. This company was organized in February, 1912, with John J. Cashin, president, and Frank L. McCaffrey, vice-president, and Edward McEwan, treasurer, and was located at 6 and 7 Long Wharf. Mr. Cashin is an old, experienced fish dealer, and has been a long time engaged with various houses. Way back in the old days he was employed with the Haskins Fish Co., then located on Commercial Wharf. He left them, after several years to go with W. J. Emerson and continued with him, coming to T Wharf in 1884, until Mr. Emerson sold his business to A. Booth & Co., and remained with this latter firm until, as stated above, in 1912 he went into business for himself together with Mr. McCaffrey and Mr. McEwan, who also worked for the Booth company for some 10 or 12 years. They remained on Long Wharf until last April, when they, with all others, joined the procession that cold, dreary 31st day of March, which led to the new fish pier, where they are now located at store No. 6. All are careful, painstaking, hard-working men and, taken together, they deserve success, which we feel sure is coming their way.

Williams Bros. Fish Co. Joseph J. Williams and Frank Williams, his brother, started in the wholesale fresh fish business on Atlantic avenue, in 1901, with perhaps the hardest outlook that often confronts two energetic and aspiring young men. The day they opened their store the price of haddock from the vessels was 7½c. per pound, running up before the close of that week to 12c. per pound, almost enough to frighten them out of their wits, but with courage and preseverance they held on, worked through the difficulties which they met, and have come to be recognized by all the trade as a house of fair dealing and good intention. Both brothers had been previously in the retail business with a store in Lynn, where Frank Williams still remains and runs the store as of old. They begun in a small way, doing good business, catering more particularly to the smaller trade, and especially the local markets. They were incorporated under the above firm name in 1910, and in April, 1911, Joseph Williams bought out his brother's interest and continued the business alone. They remained in that small store on Atlantic avenue until last April, when they came with the crowd on their pilgrimage to the New Pier, Mr. Joseph Williams having been very active in such movement. Mr. Williams is a very pleasant man to meet. He always wears a smile.

Watts & Cook Co. Many years ago, in the sixties and early seventies, the firm of Brown, Seavey & Co. were among the largest of the fish dealers in Boston, their business originally being entirely in salt fish, but as they advanced they also took on fresh fish. Like all the older houses, death and other causes have changed the firm, and some twenty-six years ago, the firm of James Emery & Co., had become successors to the first mentioned firm, and with this latter firm were employed Albert E. Watts as buyer, and Ephraim N. Cook as accountant. They were then located on Commercial Wharf. Not long after, James Emery & Co., going out of business, Mr. Watts and Mr. Cook succeeded to that firm, and carried on the business at the same old stand. They remained there until the formation of the National Fish Co., when they removed to T Wharf, taking the store formerly occupied by Baker, Wetherell & Co., and at the dissolution of the National Fish Co. they still remained on T Wharf. They have always done a large wholesale business, running more especially to the larger box trade with New York, Philadelphia and Baltimore, although catering, of course, to the barrel trade and the nearby home markets. Mr. Watts was one of the original directors and vice-president of the New England Fish Exchange for five years, and is now a director in the Boston Fish Market Corporation. They handle good stock, endeavor to get good prices and are very genial people to deal with. They also abandoned T Wharf with all the other wholesalers, and are now located on the New Fish Wharf.

Whitman, Ward & Lee Co. This house was incorporated May 22, 1911, with Chauncey W. Lee as president, Dana F. Ward as vice-president and treasurer, and Louis H. Lee, secretary. The establishment of this business, however, runs back to 1872, when Stephen Snow established himself in the wholesale fish business, on Commercial Wharf, and was located there for many years. In 1884 he came to T Wharf with all the other dealers, and in 1888 sold out to Fernald & Co., but had previously associated with himself as partner a man by the name of Krogman. Fernald & Co. continued the business at No. 29 T Wharf for about three years, when they, in turn, sold out to the late C. J. Whitman & Co. That was in the year 1891. They carried on this business at the same old stand until November, 1910, when Mr. Whitman died and Mr. Lee, mentioned above, who had been

with Mr. Whitman for many years, managed the business for the administratrix until April, 1911, when the present corporation was formed. Mr. Lee, in his early life, had always followed the sea.

Dana F. Ward had been engaged in the grocery and provision business. In June, 1900, he came to T Wharf as bookkeeper for Freeman & Cobb. In October, 1906, he entered the firm and became its manager. In 1911 he sold out his interest in the Freeman & Cobb Co., and together with Mr. Lee incorporated the business under its present style. They believe fully in system and method, with the result that the sales of the Whitman, Ward & Lee Co. have trebled since its incorporation. They do a good, substantial business, catering to the best trade throughout the New England States, also reaching as far West as Omaha, Neb., and from the principal cities of Canada to New Orleans, La.

Rush Fish Company This company succeeds the Hamilton Fish Co. Previous to 1894 Mr. Hamilton was a partner in the firm of N. D. Freeman & Co., at 24 T Wharf. In December of that year Mr. Hamilton withdrew from that partnership and started in business for himself at Nos. 8 and 9 Long Wharf. Two years later he removed to 10 and 11 Long Wharf, and remained there until the formation of the National Fish Co., when he went back to T Wharf and located at No. 6, sharing the store with O. H. Wiley & Co., until the dissolution of the National Co. Then he took store No. 14 on T Wharf. Mr. Hamilton died in 1908 and an administrator, who married Mr. Hamilton's daughter, carried on the business until May, 1913. At that time the corporation of the Hamilton-Prior Fish Co. was formed, consisting of George H. Prior as president, Mrs. Hamilton as clerk, and Wm. L. Tracy, who had been employed by Mr. Hamilton since first starting in business, as treasurer. They were capitalized at the time for $15,000, which was later increased to $20,000. In April of the past year, in company with all the wholesale fresh fish dealers, they removed to the New Fish Pier at South Boston, and are located at No. 12 at said pier at that time John F. O'Hara, president of the Star Fish Co., acquired the whole of the stock. The name of the firm was changed to the Rush Fish Co.

NOTES

Chapter 1   pages 3–21

1. Z. William Hauk, *T Wharf Notes and Sketches* (Boston: Alden-Hauk, Inc., 1952), pp. 13, 14, 21–25, 32.

2. *Fishing Gazette*, 1 May 1915, p.547; William H. Bunting, *Portrait of a Port: Boston, 1852–1914* (Cambridge, Massachusetts: Belknap Press of Harvard University Press, 1971), p. 44.

3. *Fishing Gazette*, 1 May 1915, pp. 548, 551, 552.

4. *Ibid.*, p. 552.

5. *Fishing Gazette*, 20 July 1912, p. 915; 26 October 1912; Fishing Masters' Association, *Fishermen of the Atlantic, 1914* (Boston: Fishing Masters' Association, 1914), pp. 117, 119; Frederick Roche, "The Boston Fish Pier," *International Marine Engineering* 19 (1914): 388–90.

6. Massachusetts Bureau of Statistics of Labor, *The Census of Massachusetts: 1885*, vol. 2, *Manufactures, the Fisheries and Commerce* (Boston: Wright & Potter, 1888), pp. 1411, 1430–31. U.S. Commission of Fish and Fisheries, *Report of the Commissioner for 1888* (Washington, D.C.: Government Printing Office, 1892), p. 304.

7. U.S. Commission of Fish and Fisheries, *Report of the Commissioner for the Year Ending June 30, 1898*, p. CLV.

8. *Ibid.*

9. U.S. Commission of Fish and Fisheries, *Report of the Commissioner for the Year Ending June 30, 1900*, p. 350; *Report of the Bureau of Fisheries, 1904*, p. 122.

10. U.S. Bureau of Fisheries, *Report of the Commissioner for the Fiscal Year 1908*, p. 20; U.S. Bureau of the Census, *Fisheries of the United States, 1908* (Washington, D.C.: Government Printing Office, 1911), p. 154.

11. *Fishing Gazette*, 17 November 1906, p. 1097; 1 May 1915, p. 552.

12. *Ibid.*, 1 May 1915, pp. 550–51.

13. *Boston Post*, 15 July 1908, p. 11.

14. U.S. Bureau of Fisheries, *Report of the United States Commissioner of Fisheries for the Fiscal Year 1915*, p. 58; *Fishermen of the Atlantic, 1914*, p. 119.

15. *Fishing Gazette*, 7 March 1914, p. 304; Gordon W. Thomas, *Fast & Able* (Gloucester: Gloucester 350th Anniversary Celebration, Inc., 1973), p. 111; *Fishing Gazette*, 9 April 1906, p. 329.

16. *Fishing Gazette*, 4 April 1914, p. 432; U.S. Works Projects Administration, *Boston Looks Seaward* (Boston: Bruce Humphries, 1941), p. 219.

17. *Fishing Gazette*, 27 June 1914, p. 820; 9 January 1915, p. 49; Hauk, *T Wharf*, p. 32.

18. *Fishing Gazette*, 28 February 1914, p. 275.

19. *Ibid.*, 26 November 1910, p. 1490.

20. *Ibid.*, 1 May 1915, p. 552.

21. Bunting, *Portrait of a Port*, pp. 44, 286; advertisement in *Fishermen of the Atlantic, 1909*.

22. Howard I. Chapelle, *The American Fishing Schooners, 1825–1935* (New York: W. W. Norton & Co., Inc., 1973), pp. 215–17; *Boston Post*, 4 June 1911, p. 22.

23. *Fishing Gazette*, 1 May 1915, p. 564; see Appendix 2.

24. *Boston Post*, 24 July 1911, p. 11.

25. Chapelle, *American Fishing Schooners*, pp. 167–70; Thomas, *Fast & Able*, p. 63; *Fishing Gazette*, 21 March 1908, p. 327; 23 June 1912, p. 828.

26. *Fishing Gazette*, 18 July 1908, p. 806; 19 September 1908, p. 1058; 26 December 1908; 9 January 1909, p. 20.

27. *Fishing Gazette*, 18 July 1908, p. 793; 29 August 1908, p. 965; 3 October 1908, p. 1101; 6 February 1909, p. 162; 12 February 1910, p. 162; 25 February 1911, p. 226.

28. *Fishing Gazette*, 27 March 1909, p. 404; 13 December 1913, p. 1587.

29. *Gloucester Daily Times*, 9 November 1908.

30. Harold B. Clifford, ed., *Charlie York: Maine Coast Fisherman* (Camden, Maine: International Marine Publishing Co., 1974), p. 37.

31. *Ibid.*, p. 63; *Boston Post*, 6 September 1908, p. 17; *Fishing Gazette*, 4 October 1913, p. 1273.

32. James Thomson, "The Fishing Fleet at Boston," *Harper's Weekly*, 6 December 1913, p. 22; *Fishing Gazette*, 8 June 1912, p. 723.

33. *Gloucester Daily Times*, 7 March 1911.

34. *Fishing Gazette*, 31 October 1908, p. 1223; *Commercial and Financial New England* (Boston: *Boston Herald*, 1906), p. 354; *Fishing Gazette*, 1 May 1915, pp. 548, 549.

35. Gordon W. Thomas, *Wharf & Fleet* (Gloucester: Nautical Reproductions, 1977), p. 67; *Fishing Gazette*, 9 July 1910, p. 840; 1 October 1910, p. 1257; 28 March 1908, p. 345.

36. Chapelle, *American Fishing Schooners*, pp. 200–205; *Fishing Gazette*, 13 December 1913, p. 1587.

37. *Fishing Gazette*, 27 March 1909, p. 414.

38. *Ibid.*, December 1926, p. 64.

39. *Boston Post*, 5 September 1908, p. 11.

40. *Ibid.*, 4, 5 September 1908.

41. *Fishing Gazette*, 1 May 1915, p. 548; George Brown Goode, ed., *Fisheries and Fishery Industries of the United States*, vol. 5 (Washington, D.C.: Government Printing Office, 1887), pp. 323–24.

42. Albert Cook Church, "The Evolution and Development of the American Fishing Schooner," *Yachting* 7 (1910); p. 501; see Clifford, *Charlie York*, pp. 23–45, for a description of swordfishing.

43. *Boston Post*, 20 July 1912, p. 11.

44. *Fishing Gazette*, 24 September 1910, p. 1209; 7 December 1912, p. 1559.

45. Clifford, *Charlie York*, p. 64.

46. *Fishing Gazette*, 21 February 1914, p. 243.

47. *Ibid.*, 5 December 1914, p. 1551; 27 December 1913, p. 1656; 21 December 1912, p. 1623; 25 September 1909, p. 1220.

48. *Boston Post*, 1 April 1914, p. 11; *Fishing Gazette*, 25 February 1911, p. 226.

49. *Fishing Gazette*, 27 March 1909, p. 422; 15 March 1913, p. 339.

50. *Boston Post*, 1 April 1914, p. 11.

Chapter 2 pages 22–38

1. Statement of Dana Story, Essex, Massachusetts.
2. Howard I. Chapelle, *The American Fishing Schooners, 1825–1935* (New York: W. W. Norton & Co., Inc., 1973), p. 245.
3. Dana Story, *Frame Up!* (Barre, Massachusetts: Barre Publishers, 1966), p. 124.
4. *Yachting* 12 (1912): 108–09; plans in Ships' Lines Collection, G. W. Blunt White Library, Mystic Seaport Museum, Mystic, Connecticut; *Gloucester Daily Times*, 13 March 1913.
5. Story, *Frame Up!*, pp. 117–20; Albert Cook Church, "The Evolution and Development of the American Fishing Schooner," *Yachting* 7 (1910): 503.
6. Statement of Dana Story.
7. Church, "Evolution of the American Fishing Schooner," p. 501; *Fishing Gazette*, 15 March 1913, p. 339; 9 March 1908, p. 272.
8. T. A. Scott, Inc., Papers, box 22, folder 4, G. W. Blunt White Library, Mystic Seaport Museum, Mystic, Connecticut.
9. Story, *Frame Up!*, pp. 6, 22, 48.
10. Church, "Evolution of the American Fishing Schooner," p. 502.
11. Statement of Dana Story.
12. *Fishing Gazette*, 24 October 1914, p. 1359; 20 February 1915, p. 247.
13. Statement of Dana Story.
14. Chapelle, *American Fishing Schooners*, pp. 285–89; *Fishing Gazette*, 29 September 1912, p. 1237.
15. Gordon W. Thomas, *Fast & Able* (Gloucester: Gloucester 350th Anniversary Celebration, Inc., 1973), p. 57.
16. *Ibid.*, pp. 107–08.
17. *Ibid.*, p. 57; T. A. Scott, Inc., Papers, box 13, folder 24; *Boston Commercial*, 16 March 1912, p. 3.
18. *Fishing Gazette*, 5 December 1908, p. 1358; 8 January 1916, p. 56; J. Earl Clauson, "The Price of Gasolene," *Yachting* 13 (1913): 20–21; advertisements in Fishing Masters' Association, *Fishermen of the Atlantic, 1914, 1916* (Boston: Fishing Masters' Association, 1914, 1916); Thomas, *Fast & Able*, p. 167; *International Marine Engineering* 16 (1911): 464.
19. *Fishing Gazette*, 13 June 1914, p. 761; 22 August 1914, p. 1075; Thomas, *Fast & Able*, p. 167.
20. Story, *Frame Up!*, pp. 87, 117; *Fishing Gazette*, 1 May 1915, p. 567.
21. Statement of Dana Story.
22. *Fishing Gazette*, 15 August 1914, p. 1035; 19 September 1914, p. 1209; 24 October 1914, p. 1359.
23. *Fishing Gazette*, 15 August 1914, p. 1035; Thomas, *Fast & Able*, pp. 182–83.
24. *Fishing Gazette*, 13 February 1915, p. 217.
25. Statement of Dana Story.
26. Church, "Evolution of the American Fishing Schooner," p. 501.
27. *International Marine Engineering* 17 (1912): 408; *Gloucester Daily Times*, 29 March 1913.
28. Thomas, *Fast & Able*, p. 167.
29. *Fishing Gazette*, 18 January 1913, p. 83; 29 March 1913, p. 402; 5 April 1913, p. 435; 12 April 1913, p. 467; 13 September 1913, p. 1171.
30. Story, *Frame Up!*, pp. 114–15.
31. *Fishing Gazette*, 1 February 1913, p. 146; 19 July 1913, p. 915.
32. Chapelle, *American Fishing Schooners*, p. 280; an earlier view of the *Knickerbocker's* deck can be seen in Dana Story and John A. Clayton, *The Building of a Wooden Ship: "Sawn Frames and Trunnel Fastened"* (Barre, Massachusetts: Barre Publishers, 1971), p. 77.
33. Story and Clayton, *Building of a Wooden Ship*, pp. 78–79.
34. T. A. Scott, Inc., Papers, box 23, folder 16.
35. Thomas, *Fast & Able*, p. 69.
36. Rennie Stackpole, *American Whaling in Hudson Bay* Mystic, Connecticut: Marine Historical Association, 1969), pp. 47, 63.
37. *Rudder* 16 (1905): 112.
38. *Rudder* 14 (1903): 20.

Chapter 3 pages 39–58

1. *Gloucester Daily Times*, 11 October 1912.
2. *Fishing Gazette*, 13 January 1912, p. 83.
3. *Fishing Gazette*, 4 May 1912, p. 563; 13 July 1912, p. 882; Gordon W. Thomas, *Fast & Able* (Gloucester: Gloucester 350th Anniversary Celebration, Inc., 1973), pp. 136–38.
4. T. A. Scott, Inc., Papers, box 9, folder 15; box 10, folder 18; box 23, folder 16, G. W. Blunt White Library, Mystic Seaport Museum, Mystic, Connecticut.
5. Howard I. Chapelle, *The American Fishing Schooners, 1825–1935* (New York: W. W. Norton & Co., Inc., 1973), p. 285.
6. *Gloucester Daily Times*, 25 October 1913; *Fishing Gazette*, 13 March 1915, p. 322.
7. *Fishing Gazette*, May 1927, p. 43.
8. *Ibid.*, 17 January 1914, p. 82.
9. *Ibid.*, 5 April 1913, p. 440.
10. Gordon & Hutchins Account Book, 1915–1923, pp. 8, 10, 15, 17, 40, G. W. Blunt White Library, Mystic Seaport Museum, Mystic, Connecticut; Joseph C. O'Hearn, *New England Fishing Schooners* (Milwaukee: Kalmbach Publishing Co., 1947), p. 13.
11. Gordon & Hutchins Account Book; *Fishing Gazette*, 4 October 1913, p. 1272; 9 August 1902, p. 500; 17 September 1910, p. 1160.
12. Charles H. Stevenson, *The Preservation of Fishery Products for Food* (Washington, D.C.: Government Printing Office, 1899), p. 363.
13. Wesley G. Pierce, *Goin' Fishin'* (Salem: Marine Research Society, 1934), p. 126.
14. Stevenson, *Preservation of Fishery Products*, p. 362.
15. *Fishing Gazette*, 8 November 1913, p. 1427.
16. Pat Amaral, *They Ploughed the Seas: Profiles of Azorean Master Mariners* (St. Petersburg, Florida: Valkyrie Press, Inc., 1978), p. 16; advertisements in Fishing Masters' Association, *Fishermen of the Atlantic, 1909, 1914, 1916* (Boston: Fishing Masters' Association, 1909, 1914, 1916).
17. George Brown Goode, ed., *The Fisheries and Fishery Industries of the United States*, sect. 2 (Washington, D.C.: Government Printing Office, 1887), p. 158; John R. Neal & Co., *Sea Food: A Brief History of the New England Fisheries* (Boston: John R. Neal & Co., 1893), p. 9; Pierce, *Goin' Fishin'*, p. 64; *Gloucester Daily Times*, 17 August 1914.

18. *Boston Post*, 15 July 1908, p. 11; *Fishing Gazette*, 20 March 1909, p. 373.

19. *Fishing Gazette*, 20 March 1909, p. 373.

20. *Fishing Gazette*, 17 August 1912, p. 1026.

21. *Gloucester Daily Times*, 17 August 1914.

22. Roy W. Pigeon, "A Fishing Trip to the Western Banks on the Schooner *Mayflower*," *Yachting* 31 (1922): 273.

23. *Fishing Gazette*, 20 March 1909, p. 376.

24. James B. Connolly, "The Gloucester Fisherman: Night Seining and Winter Trawling," *Scribner's Magazine* 31 (1902): 403.

25. Frederick William Wallace, *Roving Fisherman* (Gardenvale, Quebec: Canadian Fisherman, 1955), pp. 133–34.

26. *Fishing Gazette*, 4 April 1914, p. 434; *Boston Post*, 1 April 1914, p. 11.

27. Thomas, *Fast & Able*, pp. 167–69.

28. Frederick William Wallace, "The Deep Sea Trawlers," *Outing* 62 (1913): 31; A. B. Alexander, H. F. Moore, and W. C. Kendall, "Otter Trawl Fishery," *Report of the United States Commissioner of Fisheries for the Fiscal Year 1914*, Appendix 6 (Washington, D.C.: Government Printing Office, 1915), p. 17.

29. *Fishing Gazette*, 1 October 1910, p. 1246.

30. *Fishing Gazette*, 18 July 1908, p. 806; 24 October 1908, p. 1191; 1 May 1909, p. 536; 17 January 1914, p. 83; 7 February 1914, p. 179; O'Hearn, *New England Fishing Schooners*, p. 11.

31. *Fishing Gazette*, 25 May 1912, p. 629.

32. Pierce, *Goin' Fishin'*, pp. 237–41.

33. "Fifteen Days Aboard a Fishing Schooner," *The Outlook* 103 (1913): 138.

34. *Gloucester Daily Times*, 24 April 1913.

35. Taped interview with Mike Stanton, fisherman, 1966, Oral History Archives, Mystic Seaport Museum, Mystic, Connecticut.

36. *Fishing Gazette*, 18 November 1910, p. 1446.

37. *Fishermen of the Atlantic, 1913*; *Fishing Gazette*, 25 March 1906, p. 281.

38. *Fishing Gazette*, 18 May 1912, p. 627.

39. *Boston Post*, 1 April 1914, p. 11; *Gloucester Daily Times*, 25 March 1914.

40. *Report of the Commissioner of Fisheries for the Fiscal Year 1909*, p. 21.

41. *Boston Post*, 27, 28 January 1914.

42. Statement of Gordon W. Thomas; *Fishing Gazette*, 26 December 1914, p. 1647.

43. *Fishing Gazette*, July 1925, p. 22; July 1929, p. 29.

44. Goode, *Fisheries of the United States*, sect. 5, p. 270.

45. *Ibid.*, pp. 250–51.

46. *Higgins & Gifford New Catalog* (Gloucester, ca. 1893), pp. 1, 3.

47. *Gloucester Daily Times*, 15 September 1913, p. 5.

48. Gerald Fitzgerald, "Spring Mackerel Fishery," *Fishing Gazette*, 5 June 1926, pp. 12–13.

49. *Fishing Gazette*, 20 September 1913, p. 1208; 12 February 1910, p. 162; *Gloucester Daily Times*, 15 September 1913.

50. *Gloucester Daily Times*, 7 February 1914.

Chapter 4   pages 59–85

1. *Fishing Gazette*, 6 August 1910, p. 984.

2. *Ibid.*, 6 May 1912, p. 548.

3. *Ibid.*, August 1925, p. 24.

4. *Ibid.*, 18 July 1908, p. 803; 13 March 1909, p. 340.

5. *Ibid.*, 21 October 1905, p. 826; 23 January 1910, p. 84; 15 January 1910, p. 51; 20 December 1913, p. 1619.

6. "Fifteen Days Aboard a Fishing Schooner," *The Outlook* 103 (1913): 139.

7. Massachusetts Bureau of Statistics of Labor, *The Census of Massachusetts: 1885*, vol. 2, *Manufactures, the Fisheries and Commerce* (Boston: Wright & Potter, 1888), p. 1432.

8. Jeremiah Diggs, *In Great Waters* (New York: MacMillan Co., 1941), pp. 57–58; Harold A. Innis, *The Cod Fisheries: the History of an International Economy* (New Haven: Yale University Press, 1940), pp. 425–26; *The Fisherman* 5: 3 (1900): 1.

9. *Fishing Gazette*, 27 December 1913, p. 1656.

10. *Ibid.*, 3 July 1915, p. 862; May 1928, p. 19.

11. *Boston Post*, 14 January 1909, p. 13.

12. *Fishing Gazette*, 18 April 1908, p. 439.

13. See Wesley G. Pierce, *Goin' Fishin'* (Salem: Marine Research Society, 1934), and Harold G. Clifford, *Charlie York: Maine Coast Fisherman* (Camden, Maine: International Marine Publishing Co., 1974).

14. *Fishing Gazette*, 17 October 1908, p. 1223; *Boston Post*, 12 January 1909, p. 11.

15. *Ibid.*, 25 February 1911, p. 226; *Gloucester Daily Times*, 10 February 1913.

16. *Fishing Gazette*, 4 April 1908, p. 370; 2 May 1908, p. 499; Howard I. Chapelle, *The American Fishing Schooners, 1825–1935* (New York: W. W. Norton & Co., Inc., 1973), p. 280.

17. *Fishing Gazette*, 29 May 1909, p. 685; 12 February 1910, p. 162; 13 March 1915, p. 322.

18. T. A. Scott, Inc., Papers, box 10, folder 5, G. W. Blunt White Library, Mystic Seaport Museum, Mystic, Connecticut.

19. Z. William Hauk, *T Wharf Notes and Sketches* (Boston: Alden-Hauk, Inc., 1952), p. 146.

20. *Fishing Gazette*, 27 June 1908, p. 723; 18 July 1908, p. 806; 19 September 1908, p. 1058; T. A. Scott, Inc., Papers, box 22, folder 4; a two-week trip aboard the *Evelyn M. Thompson* is described in "Fifteen Days Aboard a Fishing Schooner," *The Outlook* 103 (1913): 138–40.

21. William H. Bunting, *Portrait of a Port: Boston 1852–1914* (Cambridge, Massachusetts: Belknap Press of Harvard University Press, 1971), p. 68.

22. *Boston Post*, 16 March 1908, p. 9; Gordon W. Thomas, *Fast & Able* (Gloucester: Gloucester 350th Anniversary Celebration, Inc., 1973), p. 99; *Fishing Gazette*, 9 April 1906, p. 329.

23. *Gloucester Daily Times*, 13, 14, 17 August 1912.

24. Thomas, *Fast & Able*, pp. 170–72.

25. *Fishing Gazette*, 20 April 1912, p. 503.

26. Thomas, *Fast & Able*, pp. 135–36.

27. T. A. Scott, Inc., Papers, box 11, folder 4.

28. Diggs, *In Great Waters*, pp. 57–58.

29. U.S. Commission of Fish and Fisheries, *Report of the Commissioner for the Year Ending June 30, 1900* (Washington, D.C., Government Printing Office, 1901), p. 349.

30. *Fishing Gazette*, 17 February 1906, p. 158; Fishing Masters' Association, *Fishermen of the Atlantic, 1914* (Boston: Fishing Masters' Association, 1914), pp. 133–34.

31. *Gloucester Daily Times*, 26 November 1913, p. 6.

32. Mary Heaton Vorse, "The Portuguese of Province-

town," *The Outlook* 97 (1911): 409–16; "Drudgin,'" *New England Magazine* 41 (1910): 665–71; U.S. Commission of Fish and Fisheries, *Report of the Commissioner for the Year Ending June 30, 1900*, p. 349; A. B. Alexander, H. F. Moore, and W. C. Kendall, "Otter Trawl Fishery," *Report of the United States Commissioner of Fisheries for the Fiscal Year 1914*, p. 56.

33. *Gloucester Daily Times*, 21 August 1912.

34. *Ibid.*, 20 August 1912.

35. Chapelle, *American Fishing Schooners*, pp. 167–68; *Fishing Gazette*, 23 October 1909, p. 1353.

36. *Fishing Gazette*, 31 August 1912, p. 1107; *Gloucester Daily Times*, 28 August 1912.

37. *Gloucester Daily Times*, 19 August 1912.

38. Chapelle, *American Fishing Schooners*, pp. 181–83.

39. Thomas, *Fast & Able*, pp. 22, 40, 113, 118, 183.

40. *Gloucester Daily Times*, 22 September 1908; *Fishing Gazette*, 26 December 1908, p. 1439.

41. Thomas, *Fast & Able*, p. 108; *Fishing Gazette*, 11 April 1908, p. 411.

42. *Gloucester Daily Times*, 30 June 1914.

43. George Brown Goode, ed., *The Fisheries and Fishery Industries of the United States*, sect. 5 (Washington, D.C.: Government Printing Office, 1887), p. 330.

44. *Ibid.*, p. 333.

45. U.S. Bureau of Fisheries, *Report of the United States Commissioner of Fisheries for the Fiscal Year 1914* (Washington, D.C.: Government Printing Office, 1915), pp. 18–19. For an account of the *Alaska*, see *Nautical Gazette*, 2 July 1881, p. 89.

46. *Report of the U.S. Commissioner of Fisheries for 1915*, p. 60.

47. These figures were obtained from the U.S. Treasury Department, *Annual Report of the U.S. Life-Saving Service for the Fiscal Years 1908–1914* (Washington, D.C.: Government Printing Office, 1909–1915).

48. *Fishing Gazette*, 19 February 1910, p. 201; T. A. Scott, Inc., Papers, box 13, folder 2.

49. T. A. Scott, Inc., Papers, box 7, folder 23; box 9, folder 15; box 10, folder 18.

50. *Gloucester Daily Times*, 24 July 1911.

51. Thomas, *Fast & Able*, pp. 159–61.

52. *Boston Post*, 13 July 1908, p. 9.

53. Chapelle, *American Fishing Schooners*, pp. 200–05.

54. *Gloucester Daily Times*, 24 July 1911; *Fishing Gazette*, 12 February 1912, p. 162.

55. *Fishing Gazette*, 11 November 1911, p. 1427; 18 November 1911; 25 November 1911, p. 1496.

56. See Captain W. J. Lewis Parker, *The Great Coal Schooners of New England, 1870–1909* (Mystic Connecticut: Marine Historical Association, 1948).

57. *Fishing Gazette*, 31 August 1912, p. 1107; 29 November 1913, p. 1529; 7 November 1914, p. 1419.

Chapter 5   pages 87–96

1. *Gloucester Daily Times*, 16 February 1911.

2. *Ibid.*, 9 February 1911.

3. *Ibid.*, 14 February 1911.

4. *Fishing Gazette*, 27 February 1909, p. 271; William H. Bunting, *Portrait of a Port: Boston 1852–1914* (Cambridge, Massachusetts: Belknap Press of Harvard University Press, 1971), p. 186.

5. These figures were obtained by averaging the combined monthly totals of landings in 1907, 1908, 1911, 1913, and 1914 for each three-month season. The monthly totals were found in *Report of the U.S. Commissioner of Fisheries for the Fiscal Year 1907, 1908, 1911, 1913, 1914*, tables of quantities and values of certain fishery products landed.

6. Howard I. Chapelle, *The American Fishing Schooners, 1825–1935* (New York: W. W. Norton & Co., Inc., 1973), pp. 231–34.

7. *Fishing Gazette*, 1 November 1913, p. 1395; 6 December 1913, p. 1554.

8. *Boston Post*, 5 September 1908, p. 11.

9. U.S. Bureau of Fisheries, *Report of the United States Commissioner of Fisheries for the Fiscal Year 1912, 1914, 1915* (Washington, D.C.: Government Printing Office, 1914, 1915, 1917), tables of quantities and values of certain fishery products landed at Boston and Gloucester.

10. *Fishing Gazette*, 6 August 1910, p. 979; 7 February 1914, p. 184.

11. *Fishing Gazette*, 31 October 1908, p. 1223; 30 January 1909, p. 130; 1 May 1915, pp. 549–51.

12. *Gloucester Daily Times*, 16 February 1911.

13. *Fishing Gazette*, 5 February 1910, p. 141; 21 March 1914, p. 370.

14. *Ibid.*, 27 March 1909, p. 419.

15. Fishing Masters' Association, *Fishermen of the Atlantic, 1914* (Boston: Fishing Masters' Association, 1914), p. 119.

16. *Fishing Gazette*, 20 September 1913, p. 1209.

17. Gordon W. Thomas, *Fast & Able* (Gloucester: Gloucester 350th Anniversary Celebration, Inc., 1973), pp. 148–50.

18. *Fishing Gazette*, 23 July 1911, p. 915.

19. *Fishing Gazette*, 7 March 1914, p. 290; 17 February 1912, p. 210; *Gloucester Daily Times*.

20. Thomas, *Fast & Able*, pp. 129–30.

21. *Fishing Gazette*, 3 February 1912, p. 175; 6 April 1912, p. 445; 29 September 1912, p. 1231; 1 May 1915, p. 548.

22. Donald Tressler, *Marine Products of Commerce* (New York: Chemical Catalog Co., 1923), pp. 488–89; U.S. Bureau of the Census, *Fisheries of the United States, 1908* (Washington, D.C.: Government Printing Office, 1911), p. 160.

23. Tressler, *Marine Products of Commerce*, pp. 488–89.

24. See "American and Canadian Fishing Schooners," *The American Neptune*, Pictorial Supplement 8 (1966), plate 32, the *Bessie A. Anderson*, for an example of the type.

25. *Fishing Gazette*, 25 January 1908, p. 85; 6 February 1909, p. 161.

Chapter 6   pages 97–103

1. "American and Canadian Fishing Schooners," *The American Neptune*, Pictorial Supplement 8, (1966), plate 17.

2. Gordon W. Thomas, *Fast & Able* (Gloucester: Gloucester 350th Anniversary Celebration, Inc., 1973), p. 135.

3. Fishing Masters' Association, *Fishermen of the Atlantic, 1914* (Boston: Fishing Masters' Association, 1914), p. 119; *Gloucester Daily Times*, 23, 24, 25 April 1913.

4. William A. Baker, "Fishing Under Sail in the North Atlantic," *The Atlantic World of Robert G. Albion* (Middletown, Connecticut: Wesleyan University Press, 1975), pp. 73–74; Howard I. Chapelle, *American Sailing Craft* (New York: Kennedy Brothers, Inc., 1936), pp. 111–12.

5. Howard I. Chapelle, *The American Fishing Schooners, 1825–1935* (New York: W. W. Norton & Co., Inc., 1973), pp. 252, 268–70.

6. *Fishing Gazette*, 4 September 1909, p. 1140.

7. *Ibid.*, September 1929, p. 53; 9 December 1911, p. 1555.

8. *Ibid.*, 17 January 1914, p. 83; 7 March 1914, p. 290; 13 March 1915, p. 322; 1 April 1916, p. 418.

9. Chapelle, *American Fishing Schooners*, p. 403.

10. *Fishing Gazette*, 1 April 1916, p. 418; Wesley G. Pierce, *Goin' Fishin'* (Salem: Marine Research Society, 1934) pp. 252–53.

11. *Fishing Gazette*, 20 September 1913, p. 1208; 12 February 1910, p. 162; 2 December 1911, p. 1523; *Gloucester Daily Times*, 15 September 1913.

12. *Fishing Gazette*, 16 October 1910, p. 1301; 25 November 1911, p. 1491.

13. *Fishing Gazette*, 6 December 1913, p. 1554; 13 December 1913, p. 1587.

14. Chapelle, *American Fishing Schooners*, p. 80.

15. *Fishing Gazette*, 20 March 1908, p. 365.

16. *Fishing Gazette*, 23 April 1910, p. 488; Walter Hammond, *Mutiny on the Pedro Varela* (Mystic, Connecticut: Marine Historical Association, 1956; reprint 1977).

17. *Fishing Gazette*, 4 April 1908, p. 383; 9 January 1909, p. 20.

Chapter 7   pages 104–114

1. *Fishing Gazette*, 20 January 1906, p. 49.

2. *Fishing Gazette*, 30 June 1906, p. 615.

3. *Gloucester Daily Times*.

4. Bay State Fishing Company, *Annual Report* (Boston, 1911).

5. U.S. Bureau of Fisheries, *Report of the United States Commissioner of Fisheries for the Fiscal Year 1914* (Washington, D.C.: Government Printing Office, 1915), p. 17; *Report of the Commissioner of Fisheries for 1915*, pp. 61–62.

6. *Fishing Gazette*, 3 April 1910, p. 515; 27 August 1910, p. 1083; 25 September 1911, p. 1209; *International Marine Engineering* 16 (1911): 236; 17 (1912): 19; *Gloucester Daily Times*, 10 March 1913.

7. *Fishing Gazette*, 20 January 1906, p. 49.

8. *Gloucester Daily Times*, 19 February 1914.

9. *International Marine Engineering* 17 (1912): 19.

10. See "The New Steam Trawler *Spray*," *Nautical Gazette* 70 (1906): 73–75, for a description of the *Spray* and her first voyage; Frank H. Wood, *The Story of 40 Fathom Fish* (Boston: Bay State Fishing Co., 1931), pp. 6–8.

11. *International Marine Engineering* 16 (1911): 236; 17 (1912): 19.

12. *Fishing Gazette*, 30 December 1905, p. 1081.

13. David L. Belding, "The Otter Trawl Fishery," *Fiftieth Annual Report of the Commissioners on Fisheries and Game*. Massachusetts Public Document No. 25 (Boston: Wright & Potter Printing Co., 1916), p. 86.

14. *Nautical Gazette*, 1 February 1906, p. 74; *Fishing Gazette*, 13 July 1912, pp. 882–83; 20 July 1912, p. 914; 15 January 1916, p. 83.

15. *Gloucester Daily Times*, 20 February 1914.

16. *Ibid.*, 13 February 1914.

17. *Fishing Gazette*, 7 September 1912, p. 1139; A. B. Alexander, H. F. Moore, and W. C. Kendall, "Otter Trawl Fishery," *Report of the United States Commissioner of Fisheries for the Fiscal Year 1914*, p. 25.

18. Belding, "Otter Trawl Fishery," pp. 84–85.

19. *Fishing Gazette*, 20 April 1912, pp. 481–82.

20. *Ibid.*, 27 January 1912, p. 131.

21. *Ibid.*, 18 May 1912, p. 633.

22. *Ibid.*, 20 April 1912, p. 482; 17 November 1906, p. 1097.

23. Belding, "Otter Trawl Fishery"; Alexander, Moore, and Kendall, "Otter Trawl Fishery"; *Fishing Gazette*, 30 January 1915, p. 154.

24. *Fishing Gazette*, 7 January 1911, p. 18.

25. *Boston Post*, 15 July 1911, p. 11.

26. *Fishing Gazette*, 20 January 1906, p. 49; 2 March 1912, p. 275.

27. *Nautical Gazette*, 10 April 1911, p. 13.

28. *Gloucester Daily Times*, 7 August 1914; U.S. Bureau of Fisheries, *Report of the Commissioner of Fisheries for the Fiscal Year 1915*, p. 62.

29. See Frederick William Wallace, *Roving Fisherman* (Gardenvale, Quebec: Canadian Fisherman, 1955), for a description of a cruise for menhaden in the *Long Island*.

30. Dana Story, *Frame Up!* (Barre, Massachusetts: Barre Publishers, 1966), pp. 116, 119.

31. *Fishing Gazette*, Review Number 1925, pp. 61–64.

Chapter 8   pages 115–121

1. *Goucester Daily Times*, 16 June 1914; 14 February 1914.

2. *Fishing Gazette*, 25 April 1908, p. 462.

3. Robert F. Foerster, *Italian Emigration of Our Times* (Cambridge, Massachusetts: Harvard University Press, 1924), pp. 327–29.

4. *Ibid.*, pp. 333, 340–41; *Fishing Gazette*, 5 December 1908, p. 1357; *Boston American*, 29 March 1914, p. 8B.

5. *Fishing Gazette*, 5 December 1908, p. 1357.

6. *Ibid.*, 5 December 1908, pp. 1357–59.

7. See William H. Bunting, *Portrait of a Port: Boston, 1852–1914* (Cambridge, Massachusetts: Belknap Press of Harvard University Press, 1971), pp. 214–15; Howard I. Chapelle, *The National Watercraft Collection*, 2d. ed. (Washington, D.C.: Smithsonian Institution Press, 1976), pp. 261–62.

8. *Boston Post*, 6 September 1908, p. 7.

9. *Fishing Gazette*, 5 December 1908, pp. 1358–59; *Gloucester Daily Times*, 14 February 1914.

10. *Boston Post*, 28 Juy 1912, p. 18.

11. U.S. Bureau of the Census, *Fisheries of the United States, 1908* (Washington, D.C.: Government Printing Office, 1911), p. 162.

12. *Fishing Gazette*, 1 May 1915, p. 548.

13. *Ibid.*, 10 August 1912, p. 1011.

14. *Ibid.*, 27 June 1913, p. 820; *Boston American*, 29 March 1914, p. 8B.

15. *Fishing Gazette*, 26 December 1914, p. 1637; 30 January 1915, p. 133; 6 March 1915, p. 305.

16. *Fisheries of the U.S., 1908*, pp. 156, 159, 160.

GLOSSARY

Ballooner Colloquial term for the jib topsail, the foremost of a schooner's triangular headsails, set on the jib topsail stay, running from the top of the fore-topmast to the end of the bowsprit.

Beam trawl A conical bag of netting, up to 130 feet in length, 48 to 52 feet in width at the mouth, and perhaps 3 feet high, with a heavy wooden beam affixed horizontally across the mouth to keep it open while dragging along the bottom. Adopted by fishermen of Brixham and Barking, England, shortly after 1815, apparently based on a Roman oyster dredge and a Dutch design, the beam trawl was in widespread use in Europe by 1860. Steam power was introduced to the British beam trawl fleet in 1877, and somewhat earlier in the French fleet. By 1900 the transition had been made from beam trawl to otter trawl.

Boom guy A block and tackle used on a schooner to prevent the main boom from jibing when sailing downwind. When out of use the boom guy was hung beneath the boom; when in use the after end was attached to the boom and the forward end was attached to the rail near the fore shrouds.

Brown's Bank 2,275 square miles, extending 63 miles east to west and 43 miles north to south, lying 15 miles north of Georges Bank and 70 miles southeast of Cape Sable, Nova Scotia, it varies in depth from 120 to 450 feet, with a gravelly and partly sandy bottom. This cod, haddock, halibut, pollock, and hake ground lies approximately 250 miles from Boston.

Cape North A small area 15 miles long, from 4 to 15 miles off the northern point of Cape Breton Island, it varies in depth from 390 to 600 feet, with a clay bottom. This prime cod ground lies approximately 750 miles from Gloucester and Boston.

Cape Shore Generally the area known as the Seal Island Ground, 1,250 square miles outside the 3 mile limit, lying west from Cape Sable, Nova Scotia, it varies in depth from 90 to 420 feet, with a gravelly and rocky bottom. This prime cod, haddock, and pollock ground lies approximately 280 miles from Boston. Vessels headed to the Cape Shore might also fish on Roseway and La Have Banks, lying east of Cape Sable and northeast of Brown's Bank.

Clew The lower or after corner of a sail. On many sails, a schooner's gaff topsails for example, a clew line attached to the clew was used to furl the sail.

Cod Economically the most important New England fish since the sixteenth century, this mild-tasting, firm-fleshed fish is easily preserved with salt. Among the most prolific of fish, the cod thrives in cold water north of Cape Hatteras, to 900 feet in depth, migrating inshore in winter and offshore in summer in response to changes in water temperature. A voracious feeder on any marine organism smaller than itself, the cod can weigh over 100 pounds. In 1900 the average weight of large cod caught inshore was about 35 pounds; those caught offshore 20 to 25 pounds; and of small fish, about 12 pounds. Cod livers saved by the fishermen produced oil used in tanning.

Cusk Despite good flavor and excellent suitability for boiling, the cusk was in little demand in the market before 1900 and was not widely sought by fishermen. An inhabitant of water 60 to 1,800 feet deep north of Cape Cod, the cusk is often found on rocky ledges where it eats mollusks and crustaceans. Cusk may reach a length of 3½ feet and a weight of over 25 pounds. Hook-and-line fishermen claimed that cusk wrapped their tails around rocks when caught, often destroying fishing lines.

Deep sea lead A sounding device for navigation, composed of a graduated line and a conical weight with a hollow base in which tallow is placed to bring up bottom samples for comparison with chart readings. In water less than 120 feet deep, a hand lead with a 5 to 14-pound weight was used. For deeper water, the deep sea lead had a 28 to 30-pound weight and 750 to 1,000 feet of line.

Dogfish The bane of summer fishermen, large schools of spiny dogfish, a variety of shark, often roam the fishing grounds to depths of 600 feet, eating fish, squid, worms, crabs, and bait. Dogfish range up to 4 feet in length and 20 pounds in weight, but most are under 3½ feet and 10 pounds. Reportedly, one summer day a 500-hook trawl took one haddock, 2 cod, and 497 dogfish in a single set.

Gasket A short line for lashing a sail when furling it. Sometimes called a sail stop.

Georges Bank 11,000 square miles, 8,498 in the "winter fishing-ground," extending 150 miles northeast to southwest and 90 miles north to south, with deadly shoals in the northwest sector, lying approximately 90 miles east of Cape Cod, it varies in depth from 12 to 300 feet, with a mostly sandy bottom. Originally called St. George's Bank, this prime cod, haddock, and halibut ground lies approximately 175 miles from Boston.

Gill net A wall of netting with mesh large enough to allow fish to swim part way through. When the fish try to back out of the net, the mesh catches behind the gills, snaring the fish. Gill nets can be set on or beneath the surface.

Groundfish A general term for marketable bottom-dwelling fish. In 1910 groundfish connoted cod, haddock, hake, cusk, pollock, and halibut.

Gurry box A large box on the deck of a fishing schooner, just forward of the after cabin trunk. Until the 1880s, fishermen feared that gurry (fish scraps) thrown overboard would drive fish from the fishing grounds, so the gurry box was devised to hold gurry until a vessel left the grounds. By the twentieth century the gurry box was only used as a storage area for gear. Its top was often used for baiting.

Haddock A close relation to the cod, the haddock became Boston's most important fish before 1900. The haddock fishery had been insignificant before the rise of the fresh fishery around 1860, though the moist, tasty haddock was long known to be good fresh, boiled, or in chowders. John R. Neal & Co. established smokehouses in Boston to produce popular "finnan haddie" (smoked haddock) in the 1880s. Distinguishable from the cod by its smaller size (averaging 3 to 4 pounds and ranging to about 17) black lateral line along its side (the cod has a white lateral line) and its pointed first dorsal fin, haddock inhabit a narrower range, generally north of 40° N. Swimming in schools on bottom (especially clam banks) between 150 and 600 feet deep, haddock are omniverous and more voracious than cod; some fishermen claimed that haddock would bite hooks lying on the bottom while cod would only bite at hooks hanging slightly above the bottom. Not as prolific as cod, the haddock population fluctuates periodically.

Hake A very abundant, easily salted fish like cod, the hake was commonly corned (pickled in brine). With the development of processed "boneless cod" in the 1870s, hake was often packaged or mixed in as boneless or shredded cod. Hake liver oil was frequently used as cod liver oil, and the hake sound (swim bladder) was an important source of isinglass. Hake seem to prefer muddy bottom, at depths to 1,800 feet. Handline fishermen found they were more likely to take bait at night in depths from 60 to 300 feet. Averaging 5 to 10 pounds in weight, hake seem to feed largely on crustaceans.

Halibut Known for its firm, mild flesh for centuries, the halibut supported a valuable nineteenth-century New England fishery until overfishing depleted the stock. Thereafter, Pacific halibut met a large part of New England's demand. Inhabitants of cold water, with a range similar to that of cod, halibut are found on the bottom from inshore to a depth of 1,500 feet. Largest of the flat fish, with eyes usually on the right side of the head, halibut lie on the bottom to capture crabs, mollusks, and fish. However, they are also known to chase their prey to the surface, and have been seen to stun fish with blows of their tails. Fishermen preferred female halibut, which could reach over 400 pounds in weight (males usually weigh less than 100). A 350-pound halibut would be 7 to 8 feet long and 4 feet wide; the usual weight was less than 250 pounds. The most marketable was a female of 80 pounds, but "chicken halibut" of 10 to 20 pounds was also in steady demand. As a clean white underside brought the best price at market, fishermen scrubbed all blood out of dressed halibut and packed them upside down in the hold so the bellies did not discolor from settling fluids. Large halibut were often "greys," with blotched bellies, and brought a lower price.

Handline A fishing line composed of a long line attached to a sinker, beneath which generally hung 2 short lines, each with a baited hook. A fisherman generally tended one to 3 handlines at a time.

Herring Also called sea herring, this surface-schooling fish is found in open coastal waters from Block Island to Labrador, seeking plankton for food. The New England fishery reached its height in Maine. Herring may reach 18 inches in length, but economically the most important have been the small, 3 to 5-inch fish, packed as "sardines" since the 1870s. Large herring were packed fresh, pickled, or smoked. Herring was also important as bait, either fresh or frozen.

Indian Header Colloquial term for a model of fishing schooner having a well-rounded, convex or spoon bow and cutaway forefoot, first designed by Thomas F. McManus in 1898. The first vessels of this type had Indian names, hence the designation.

Jeffrey's Ledge A small ground of about 60 square miles, considered to extend 20 miles northeasterly from the northeast shoals of Cape Ann, and 2 to 4 miles in width, it varies in depth from 162 feet on the ledge to 300 feet at the edges, with a rocky and gravelly bottom. Resorted to by some of the small vessels, this cod, haddock, and cusk ground lies approximately 50 miles from Boston.

Jumbo Colloquial term for the forestaysail, the aftermost of a schooner's triangular headsails, set on the forestay running from the head of the foremast to the area of the stemhead. The foot of the jumbo is usually affixed to a boom.

Knockabout Colloquial term for a schooner without a bowsprit, apparently named for small Massachusetts racing sloops of the 1890s, without bowsprits, called knockabouts. Designed by Thomas F. McManus, the first knockabout fishing schooner appeared in 1902.

Lazy jacks Colloquial term for a set of vertical lines in pairs, often between the topping lift and boom of a gaff-rigged sail, that contained the folds of the sail as the gaff was dropped for furling. Generally in use on vessels with small crews, lazy jacks were rarely carried by fishing schooners.

Lemon sole Colloquial term for the winter flounder, or flatfish, not related to the soles of the eastern Atlantic. This firm, white, flavorful fish, found abundantly from the waters of New York north, has been featured in the New York market since colonial times. With eyes on the right side of its head, the fish lies on muddy to firm bottom in 6 to 120 or more feet of water eating mollusks and crustaceans, and scavenging. The average weight is about 1½ pounds, with a length of 12 to 15 inches. Migrating according to water temperature, toward the south the fish is generally found inshore during winter only, and therefore has been named the winter flounder.

Mackerel Second only to the cod in economic importance into the nineteenth century, the mackerel supported a major New England fishery from April to November. A surface schooling fish, the mackerel appears off Cape Hatteras about April and arrives off New England and Nova Scotia about June, returning south in the fall. Although they travel in schools of hundreds of thousands of fish, seeking small crustaceans and fish spawn for food, the population fluctuates from year to year, and the success of the fishery has always been unpredictable. From colonial times mackerel packing has been strictly regulated. Categories included: No. 1, best quality, over 13 inches; No. 2, best quality, 11 to 13 inches; several grades of No. 3 covering second quality fish and those under 11 inches. Though reaching 18 inches in length, the average mackerel taken was about 12 inches long and a pound in weight. Tinker mackerel were about 2 years old and under 10 inches. The choicest packaged mackerel were called "bloaters."

Market fishery An unspecialized form of fishing generally pursued by smaller vessels, making frequent short trips to deliver mixed fresh groundfish of the season to market.

Middle Bank Also called Stellwagen Bank, about 70 square miles, extending 17 miles from 15 miles south of Cape Ann to within 5½ miles of Race Point, Provincetown, and averaging 4½ miles east to west, it varies in depth from 57 to 104 feet, with a sandy and gravelly bottom. This good cod and haddock ground lies approximately 30 miles from Boston.

Misstay Or miss stays, to fail to swing the bow of a sailing vessel through the wind when going from one tack to the other. The vessel either comes to a stop, "in irons," or falls off on the original tack.

Otter trawl A conical net bag similar to a beam trawl, except having a rectangular wooden "door" or otter board where the towing lines attach at each side of the net's mouth. As the net is dragged, the doors "swim" outward like kites, keeping the mouth open. The otter trawl was apparently developed in England about 1894.

Pollock Though related to the cod and found at depths to 600 feet, pollock are generally surface swimmers, schooling in search of young fish for food. Pollock flesh is darker than cod, but actually salts better and has a pleasing taste. Slack-salted pollock was a delicacy for some fishermen; others refused to eat pollock and considered it as much a nuisance as the dogfish because it devoured the spawn of fish more valuable than itself. However, it was pursued, with both lines and purse seines, for its flesh and the oil of its liver. Pollock may reach a length of 3 feet and a weight of 25 pounds.

Purse seine A surface fishing device typically composed of a net approximately 1,200 feet long by 130 feet deep, with a drawstring along the bottom edge. Set around a school of fish on the surface, the seine was pursed up to form a net bag containing the fish.

Purser A schooner fisherman selected by the captain to oversee the vessel's accounts during a voyage, and to represent him if necessary, acting in effect as mate of the vessel.

Salt banking Colloquial term for the cod fishery pursued by vessels with salt for preservation. Salt bankers generally made one to 6 trips of several months duration between March and October, using either handlines or trawl lines. Those fishing on Western Bank made up to 6 trips; those going to the Grand Bank made one or 2 trips. The cod were split and packed in salt, about a pound of salt for every 4 pounds of fish.

Scrod In 1910 market terminology, cod or haddock weighing one to 2½ pounds. In the nineteenth century, along the eastern coast of Massachusetts, scrod referred to small cod, lightly corned (pickled in brine) and to the small fish themselves.

Semi-knockabout Colloquial term for a schooner design combining elements of Indian Header and knockabout fishing schooner types, with an elongated spoon bow, a short spike bowsprit, and the forestay brought down inboard of the head of the stem, rather than to the stemhead as in an Indian Header.

Shack fishing Colloquial term for a fishery using both ice and salt for preservation. Fish caught with the first supply of bait were salted. With a fresh supply of bait, fish caught thereafter were preserved in ice.

South Channel Generally, the area including the Great South Channel and the various shoals east of Nantucket, about 2,400 square miles, extending 40 miles east from Nantucket to Georges Bank and 60 miles south from the latitude of Chatham Light, it varies in depth from 9 to 300 feet, with a mostly sandy bottom. This prime haddock ground lies approximately 100 miles from Boston.

Splitters Colloquial term for the fish processors in Gloucester who purchased fresh groundfish and split and salted them. If prices were low in Boston, cargoes of fresh fish were often sold to the splitters.

Stock The monetary return to a vessel for its cargo of fish, the gross stock was the total amount, equal to the sum of the bid price for each species multiplied by the total weight of each species as landed on the wharf. The net stock for Boston vessels usually equalled three-quarters of the gross stock minus general expenses and the captain's percentage. It was divided equally among the crew, captain, and cook after expenses shared by the crew were deducted. In some fisheries the net stock was allotted "by the count," based on each man's catch during the trip.

Swordfish Known as a food fish at least since the seventeenth century, the swordfish became the quarry of an active New England fishery between 1840 and 1855. Found throughout the Atlantic, swordfish usually arrive off Southern New England about the first of June, spreading north, and disappearing about November. They travel singly, preying on schools of menhaden, mackerel, herring, and other surface fish, though occasionally they have been taken on trawls set on the bottom. The sword is used to kill or stun fish; it was also wielded ferociously against fishermen. Swordfish were usually found swimming slowly or lolling in calm water before 10:00 A.M. and after 4:00 P.M. Because of the danger, and because the fish were hunted singly with a harpoon-like "lily iron," this was the most sporting of the New England fisheries. It was also lucrative as the average fish weighed perhaps 525 pounds alive, and 250 pounds dressed for market.

Trap Also called pound, this stationary net structure is supported with poles or floats, and anchors. A trap usually comprised a leader running out from shore to intercept fish swimming alongshore, wings to funnel the fish, and a net enclosure to contain the fish. Pounds usually had two enclosures, an outer heart-shaped "heart" and an inner pound or bowl. Set in the spring, and often fished through summer and fall, traps and pounds were tended by fishermen at low tide.

Trawl line A form of hook-and-line fishing, traceable to Dutch fisheries, ca. 1700, allowing a single fisherman to tend many hooks. The trawl was composed of a groundline, short lines (gangings) that attached the hooks to the groundline, an anchor at either end of the groundline to hold it at the bottom, and a buoy at either end to mark its location. While the groundline was commonly about 1,800 feet long, dimensions of the other components varied from fishery to fishery, and over time.

Weir A term sometimes used interchangeably with trap or pound, but more properly used to describe a structure of brush or netting set in shallow water, often in a channel or waterway, to divert and entrap fish.

Western Bank Also called Sable Island Bank, 7,000 square miles, extending 156 miles northeast to west and 76 miles north to south, generally lying northwest and west of treacherous Sable Island, the bank varies in depth from 100 to 360 feet, with a sandy bottom. This important cod, haddock, and halibut ground lies approximately 500 miles from Boston. To the northeast, Western Bank is separated from Banquereau by the Gully, a major halibut ground 69 miles long by about 20 miles wide.

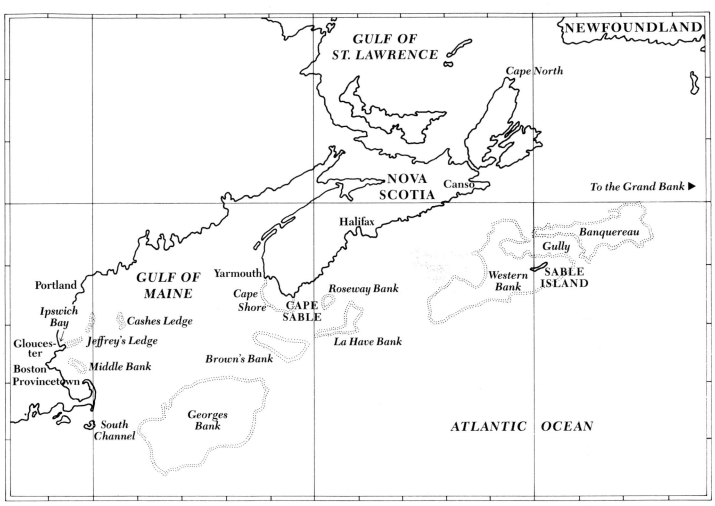

Major Fishing Grounds of the Boston Fleet, ca. 1910, adapted by Bill Gill from "Fishing Banks, Cape Cod to Grand Bank," by T. Van de Bogart in *Report of the U.S. Commissioner of Fisheries for the Fiscal Year 1914*

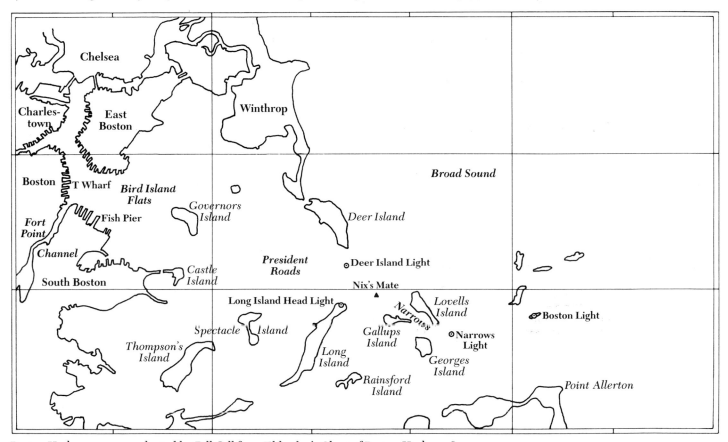

Boston Harbor, ca. 1914, adapted by Bill Gill from Eldredge's Chart of Boston Harbor, 1899

SOURCES ON THE BOSTON FISHERIES

Boston and T Wharf

"Boston as a Fishing Port—Past and Present." *Fishing Gazette*, 1 May 1915, pp. 545–69. An excellent contemporary analysis of the history of the Boston fisheries, including brief histories of the wholesale fish dealerships.

Bunting, William. *Portrait of a Port: Boston, 1852–1914.* Cambridge: Belknap Press of Harvard University Press, 1971. This exemplary photographic work presents the variety of Boston's maritime activities, including the fisheries.

Hauk, Z. William. *T Wharf Notes and Sketches.* Boston: Alden-Hauk, Inc., 1952. This privately produced notebook of T Wharf history provides the most detailed picture of T Wharf before the fishing industry located there and after it became an art colony in the 1920s.

Roche, Frederick. "The Boston Fish Pier." *International Marine Engineering* 19 (1914): 388–90. A description of the new South Boston fish pier.

Thomson, James. "The Fishing Fleet at Boston." *Harper's Weekly* 58 (6 December 1913): pp. 22–23. Thomson takes the reader on a brief walk along T Wharf.

U.S. Works Projects Administration. *Boston Looks Seaward.* Boston: Bruce Humphries, Inc., 1941. Reprint, New York: Library Editions, Ltd., 1970. The Boston Port Authority sponsored this maritime history of Boston, 1630–1940.

History of the New England Fisheries

Albion, Robert, G., Baker, William A., and Labaree, Benjamin W. *New England and the Sea.* Middletown, Connecticut: Wesleyan University Press, 1972. This excellent introductory text puts the fisheries into the perspective of New England maritime history.

Goode, George Brown, ed. *Fisheries and Fishery Industries of the United States*, 5 sect. in 7 vols. Washington, D.C.: Government Printing Office, 1887. This classic study presents the history and technology of all the significant American fisheries pursued in the 1880s. It is indispensable.

Innis, Harold A. *The Cod Fisheries: The History of an International Economy.* New Haven: Yale University Press, 1940. Innis's microscopic study of the Canadian fisheries points out the close ties across the border that developed through the nineteenth century.

McFarland, Raymond. *History of the New England Fisheries.* Philadelphia: University of Pennsylvania Press, 1911. McFarland's study is especially useful for its coverage of the pre-colonial fisheries and the diplomatic history of the New England fisheries.

Pierce, Wesley G. *Goin' Fishin'.* Salem: Marine Research Society, 1934. Pierce's obvious experience in the fisheries makes up for his lack of sources and his occasional dependence on legend.

Fishermen

Clifford, Harold B., ed. *Charlie York: Maine Coast Fisherman.* Camden, Maine: International Marine Publishing Co., 1974. This compilation of an oral history project gives excellent insight into the life of a shore fisherman. Charlie York speaks for many of the men seen in these photographs.

Conlan, Marcus, "Bald Dutch and the Silly Fish." *Yachting* 11 (1912): 96–98. A whimsical story which gives a good verbal description of swordfishing.

Connolly, James B. *The Book of the Gloucester Fishermen.* New York: John Day Co., 1930. An excellent collection of short stories by the fishermen's greatest literary proponent.

Diggs, Jeremiah. *In Great Waters.* New York: MacMillan Co., 1941. History and legend of the Portuguese fishermen of Provincetown.

"Drudgin'." *New England Magazine* 41 (1910): 665–71. Description of flounder dragging from Provincetown which gives a glimpse of the shore fisherman's life.

Kipling, Rudyard. *Captains Courageous.* First edition 1896. A boy learns about life with Gloucester fishermen on a dory handlining trip in the 1890s.

Spectator. "Fifteen Days Aboard a Fishing Schooner." *The Outlook* 100 (1913): 137–39. A journalist describes the fisherman's life and speculates on the nature of his existence.

Thompson, Winfield M. "The Passing of the New England Fisherman." *New England Magazine* 13 n.s. (1896): 675–86. A description of the "old" way of fishing in Maine and a discussion of the changes which took place late in the nineteenth century.

Vorse, Mary Heaton. "The Portuguese of Province-town." *The Outlook* 97 (1911): 409–16. A noted Provincetown author provides a good account of the Portuguese fishermen of her town.

Wasson, George S. "On the Edges of the Sea." *The Outlook* 84 (1906): 990–1003. A good description of the varieties of maritime labor, especially fishing, along the New England coast.

Dory Fishing

Campbell, Charles A. "The *Mermaid*'s Nursling." *New England Magazine* 41 (1909): 470–78. A fictitious trip on a fishing schooner, with good illustrations.

Church, Albert Cook. *The American Fishermen*. New York: W. W. Norton, 1941. This has been the best collection of fishing photographs for forty years. Many of Church's views were taken during the same period Fisher took his.

Connolly, James B. "The Gloucester Fishermen: Night Seining and Winter Trawling." *Scribner's Magazine* 31 (1902): 387–407. An excellent description. Connolly's writing is well complemented by Milton Burns's illustrations.

Goldie, George S. "A Winter Fishing Trip to George's Shoal." *Yachting* 9 (1910): 346–50, 431–36. Goldie's excellent account gives many technical details of dory trawling in winter. His photographs add much to the article.

Henderson, William J. "The Catching of the Cod." *The Century Magazine* 72 (1906): 485–96. This piece is best for its explanation of the process of finding fish and of finding a lost dory. Illustrated by Milton Burns.

Pigeon, Roy W. "A Fishing Trip to the Western Banks on the schooner *Mayflower*." *Yachting* 31 (1922): 196, 273. A description of racing and fishing in New England schooners.

Sea Food: A Brief History of the New England Fisheries. Boston: A. T. Bliss & Co., 1893. John R. Neal & Co. prepared this informative pamphlet for their booth at the Chicago Columbian Exposition of 1893. It contains useful details of various fisheries and especially on market fishing.

Wallace, Frederick W. "The Deep Sea Trawlers." *Outing* (1913): 26–35. A fine early article by a well-known authority on the fisheries. Includes good photographs by the author.

———. "Life on the Grand Banks." *The National Geographic Magazine* 40 (1921): 1–28. An expansion of the last article, with more excellent photographs.

———. *Roving Fisherman*. Gardenvale, Que.: Canadian Fisherman, 1955. A retrospective memoir of experiences on various types of fishing vessels, 1911–24. The sections on haddock and halibut trawling are the best accounts of the types of fishing pursued by the craft pictured in this book. The sections on red snapper fishing and Pacific fishing present some of the alternatives followed by New England fishermen and fishing schooners.

Mackerel Fishing

Church, Albert Cook. "The Catching of the Mackerel." *Yachting* 10 (1911): 277–280. A useful introduction to mackerel seining, with good Church photographs.

Connolly, James B. *The Seiners*. New York: Charles Scribner's Sons, 1904. The best fictional account of mackerel seining, particularly good for dialogue.

"A Cruise for Mackerel." *Fishing Gazette*, 22 November 1908, pp. 1293–94. A good description of the process of setting and hauling a seine.

Fitzgerald, Gerald. "Spring Mackerel Fishing." *Fishing Gazette*, 6 June 1926. A technical explanation of mackerel seining and mackerel gill netting.

McFarland, Raymond *Masts of Gloucester*. New York: W. W. Norton & Co., 1937. Fine description of a summer on a mackerel seiner in the 1890s.

Otter Trawling

Alexander, A. B., Moore, H. F., and Kendall, W. C. "Otter-Trawl Fishery." *Report of the U.S. Commissioner of Fisheries for 1914*. Appendix VI. Washington, D.C.: Government Printing Office, 1915. This is the report of the official government investigation of otter trawling.

Belding, David L. "The Otter Trawl Fishery." *Fiftieth Annual Report of the Commissioners of Fisheries and Game*. Massachusetts Public Document No. 25. Boston: Wright & Potter Printing Co., 1915. The first scientific study of American otter trawling.

"The New Steam Trawler *Spray*." *The Nautical Gazette*, 1 February 1906, pp. 73–75. A fine description of the *Spray* and her first fishing trip.

"Otter Trawling in Atlantic Fisheries." *Fishing Gazette*, Review Number 1925, pp. 61–64. This article outlines the first twenty years of New England otter trawling.

Symonds, Ralph F., and Trowbridge, Henry O. "The Development of Beam Trawling in the North Atlantic." *Transactions of the Society of Naval Architects and Marine Engineers*. Vol. 55. New York: Society of Naval Architects and Marine Engineers,

1948. An analysis of otter trawling vessels and equipment, with an emphasis on developments in the 1920s and '30s. Includes plans of the *Spray* and diagrams of otter trawls.

Fishing Vessels

Baker, William A. "Fishing Under Sail in the North Atlantic." In *The Atlantic World of Robert G. Albion.* Benjamin Labaree, ed. Middletown, Conn.: Wesleyan University Press, 1975. Baker gives a useful overview of vessel types and fishing methods in the North Atlantic since the 1400s.

Barnes, Albert M., ed. "American and Canadian Fishing Schooners." *The American Neptune,* Pictorial Supplement 8, 1966. A brief survey of fishing schooner types depicted in seldom-seen photographs.

Chapelle, Howard I. *The American Fishing Schooner, 1825–1935.* New York: W. W. Norton, 1973. Chapelle, the preeminent analyst of historic vessel designs, here traces fishing schooner development through existing plans and models. An appendix catalogs details observed on schooners in the 1930s. Chapelle describes several vessels seen in Fisher's photographs.

———. *American Sailing Craft.* New York: Kennedy Brothers, Inc., 1936. In this early work Chapelle presents his preliminary research on fishing schooners, sloop boats, Friendship sloops, and Cape Cod catboats, each of which was used for fishing.

———. *The National Watercraft Collection.* 2d. ed. Washington, D.C.: Smithsonian Institution Press, 1976. This work contains both a valuable essay on fishing craft and descriptions of models in the National Museum.

Church, Albert Cook. "The Evolution and Development of the American Fishing Schooner." *Yachting* 7 (1910): 409, 499. This piece contains some unsubstantiated legends, but the information on schooners of the 1880–1910 period is very useful. Good illustrations are included.

Fishermen of the Atlantic. Boston: Fishing Masters' Association, 1909, 1911, 1913, 1914, 1916. These yearbooks, when they can be found, provide registry lists for each port, data, occasional stories, and period advertisements.

"Fishing Schooner *Rob Roy* Sailing out of Gloucester, Mass." *International Marine Engineering* 6 (1901): 185–89. A compilation of data on Crowninshield's influential *Rob Roy* design. Very complete specifications are included, along with lines, construction, and sail plans.

"A Flying Fisherman." *The Rudder* 11 (1900): 178–82. A preliminary description of the *Helen Miller Gould,* the first gasoline auxiliary fishing schooner.

"The Knockabout Fishermen." *International Marine Engineering* 7 (1902): 299–301. A useful description of the *Helen B. Thomas,* the first knockabout fishing schooner. Includes photos and lines.

O'Hearn, Joseph. *New England Fishing Schooners.* Milwaukee: Kalmbach Publishing Co., 1947. A picture book which contains some seldom-seen photos and useful details on the schooners.

"The Otter Trawler *Spray.*" *Fishing Gazette,* 20 January 1906, pp. 49. The Spray was front page news for the *Fishing Gazette* in January 1906. The article briefly introduces the vessel and the fishing method to the fishing industry.

"Sicilian Motorboat Fishing Fleet of Boston." *Fishing Gazette,* 5 December 1908, pp. 1357–59. The fullest description of these gasoline boats and their uses.

"Steam Trawler *Surf* and *Swell.*" *International Marine Engineering* 17 (1912): 19. This brief article offers the best description of the layout and specifications of the early otter trawlers.

Story, Dana. *Frame Up!* Barre, Mass.: Barre Publishers, 1966. A valuable and engaging look at life and shipbuilding in Essex, by the son of the town's most prolific shipbuilder.

Story, Dana, and Clayton, John A. *The Building of a Wooden Ship: "Sawn Frames and Trunnel Fastened."* Barre, Mass.: Barre Publishers, 1971. An excellent photographic record of shipbuilding process at Essex. A few photographs date from the period covered by Henry Fisher's photographs.

Thomas, Gordon. *Fast & Able.* Gloucester: Gloucester 350th Anniversary Celebration, Inc., 1973. "Biographies" of seventy-seven fishing vessels, 1874–1930, and of some of Gloucester's greatest skippers. *Fast & Able* has been both a reference and a model for parts of this book.

———. *Wharf and Fleet.* Gloucester: Nautical Reproductions of Gloucester, 1977. A collection of eighty-one photos of Gloucester and its vessels, printed in large format. Many are contemporaneous with Fisher's photos.

Fish

Goode, George Brown. *American Fishes: A Popular Treatise Upon the Game and Food Fishes of North America.* Revised by Theodore Gill. New Edition. Boston: L. C. Page & Co., 1903. Originally published in 1887, this is an excellent compilation of natural history, lore, and practical knowledge on American fish.

———. *The Fisheries and Fishery Industries of the United States,* 5 sect. in 7 vols. Washington, D.C.: Government Printing Office, 1887. This compendium has several sections on fish species and their habitats.

Jensen, Albert C. *The Cod.* New York: Thomas Y. Crowell, 1972. This study, intended for general readers, discusses both the natural and economic histories of the cod.

Stevenson, Charles H. *Preservation of Fishery Products for Food.* Washington, D.C.: Government Printing Office, 1899. This work presents the state of the art in fish processing and preservation just before Fisher made this record of the Boston fisheries.

Tressler, Donald. *Marine Products of Commerce.* New York: Chemical Catalog Co., 1923. A scientific discussion of the processing and uses of fishery products.

U.S. Bureau of the Census. *Fisheries of the U.S., 1908.* Washington, D.C.: Government Printing Office, 1911. A useful statistical accounting of the fisheries, broken down by state.

U.S. Commissioner of Fish and Fisheries. *Annual Reports,* 1875–1920. Washington, D.C.: Government Printing Office. Title varies. A wealth of statistics and information on many fish species and fisheries can be found in these reports. The 1920 volume includes "An Analytical Subject Bibliography of the Publications of the Bureau of Fisheries, 1871–1920."

Miscellaneous

NEWSPAPERS

Boston Post. The column "Along the Waterfront" in this daily paper reports the landings of the fishing fleet and any newsworthy occurrences among the vessels.

Fishing Gazette. This weekly organ of the industry offers a tremendous amount of detail on fishing activities. The *Fishing Gazette* became a monthly after 1919. Some issues from the 1920s with retrospective articles were also used.

Gloucester Daily Times. Produced daily, except Sunday, this paper has much information on the local fisheries, including Boston's. Its details of vessel movements are somewhat more useful than those in the *Boston Post.*

Nautical Gazette. New York, weekly. This weekly magazine notes significant maritime news, describes new vessels, and reports on harbor activity, including that at Boston.

OTHER

American Bureau of Shipping. *Record of American and Foreign Shipping.* New York, yearly.

Lloyd's Register of Shipping. London, yearly.

Mercantile Navy List and Maritime Directory. Liverpool, 1914, 1925, 1940.

U.S. Department of Commerce and Labor. *Annual List of Merchant Vessels of the U.S.* Washington, D.C.: Government Printing Office, yearly.

Each of the above was used to trace vessels identified in the photographs. *Merchant Vessels of the U.S.* was most useful for fishing vessels, since they were rarely insured by, or built to specifications of, Lloyd's of London. Through the yearly listings, change in port of registry, addition of auxiliary power, and vessel sale or loss can be determined.

U.S. Treasury Department. *Annual Report of the U.S. Life-Saving Service.* Washington, D.C.: Government Printing Office, yearly. The reports record losses of vessels and individuals yearly.

INDEX